An Introduction to Elementary Computer and Compiler Design

An Introduction to Elementary Computer and Compiler Design

Dennis R. Steele

Department of Mathematics and Computer Science
Graceland College

ELSEVIER NORTH-HOLLAND, INC.
52 Vanderbilt Avenue, New York, New York 10017

Distributors outside the United States and Canada:
THOMOND BOOKS
(A Division of Elsevier/North-Holland Scientific
Publishers, Ltd.)
P.O. Box 85
Limerick, Ireland

Library of Congress Cataloging in Publication Data
Steele, Dennis R
An introduction to elementary computer and compiler design.

Bibliography: p.
Includes index.
1. Electronic digital computers--Design and construction. 2. Electronic digital computers--Programming. 3. Compiling (Electronic computers)
I. Title.
TK7888.3.S72 621.3819'58'2 77-17872
ISBN 0-444-00243-X

Manufactured in the United States of America

Writing and research cannot always be done at the office. In my case it takes a special kind of wife and cooperative children in order that such activities can be done at home.

To Linda, my wife, and to
Heather, Chad, Shannan, and Gabriel, my children,
I dedicate this book

Contents

Preface

The field of study known by most academicians as Computer Science is still in its infancy. The phenomena of the computer, however, is a bit older. Perhaps one could characterize it as a "pre-teenager"—that stage of life, at least in America, when nothing stays "settled" for more than a few days.

Developing a pedagogically sound approach to the computer is a process that may never settle in the sense that other disciplines, such as mathematics (at the college level), have settled. But one thing is certain—distinctions such as hardware-firmware-software, and techniques such as flow charting and high-level language coding will, in one form or another, be around for some time. Educators merely reflect these phenomena in their teaching and hope to find organizational schemes to make them palatable to the student. The following text is an effort in that direction.

The text is devoted exclusively to elementary hardware design and elementary compiler design. Techniques are introduced only as they are needed and therefore one will not find an overview of all similar techniques. One will find, however, a concise introduction to one of the most sophisticated forms of programming, namely, the design of a compiler for a high-level programming language. One will also find a direct connection between the compiler and the hardware, and as a result, obtain a functional overview of an elementary machine and language which, hopefully, will provide a sound foundation for further study.

The author is indebted to a number of individuals. Dr. Roger Camp of Iowa State University who required Ph.D. candidates in Electrical Engineering and Computer Science to examine a simple machine in spite of their advanced level of education. His simple machine, upon which the machine designed in the text is based, opened the eyes of many bewildered Ph.D. candidates. I shall always remember the lesson learned, namely, do not attempt advanced subjects until the basic material for those subjects is mastered.

Drs. Keller and Maple, also of Iowa State University, provided the author with considerable motivation in the area of Computer Science. Their grasp, not only of this complex field of study, but also of the need to approach the teaching of such a discipline in new and innovative ways, has provided a constant resource for the author.

The text is dedicated to these fine university professors.

Introduction

The so-called "knowledge explosion" has affected nearly every aspect of society. Only the most remote sections of the planet have managed to avoid its impact, and even in those places there is an indirect effect. No doubt our institutions of higher learning are the focal point of this phenomenon. They, together with industry, have produced the knowledge and at the same time are faced with the constant responsibility and task of keeping current.

Within the academic environment certain disciplines have felt this impact more dramatically than others. Certainly Computer Science is one of those disciplines. By the time an advancement in technology is sufficiently understood to be organized into a form conducive to a learning environment, it is usually dated and, in some cases, obsolete. Computer scientists have attempted to distill basic principles appropriate to their discipline and in many instances have done so with much success. They have been greatly assisted by the fact that their field draws heavily upon such well established disciplines as electrical engineering, linguistics, and mathematics. But even with this help there is much to be done that is a result of individual creativity and imagination. In fact, one of the classic reference volumes, *The Art of Computer Programming* by D.E. Knuth, reflects the "art-orientation" which is still necessary for anyone who wishes to do substantial work in the field. This orientation compounds even further the task of keeping current.

While educators have struggled with problems peculiar to their task, the industrial establishment has attempted to meet its challenges with equal success. One challenge that the computer industry has faced and met with tremendous results has been the challenge of the market place. Computer products and services have been marketed with notable success, and as a result are bringing about a slow but sure transformation of our society.

One particular development stands out. If the computer industry had expected all of its customers to use the computer without the aid of high-level computer languages (i.e., languages that are relatively easy for the customer to use), the marketing of computer services would have been a difficult if not impossible task.

In some high-level languages the "sentences" are nearly identical to the everyday language of the user. These languages have made an incredibly complicated machine as easy to use as some of the more complicated children's toys.

We come now to the primary reason for the following text. The "distilling" efforts of the academy have produced some principle-based educational material. The "marketing" efforts of industry have produced such things as high-level programming languages. These two have combined to produce for most students of Computer Science what I shall call the "black box" problem. A few well-applied principles of programming will allow a student to produce a program in a high-level language, execute the program on a computer, and obtain results, usually on paper, which are easy to read and use. What happened "in the middle" might be regarded as a magician's trick as far as the student is concerned. What went on inside the "black box" is a mystery.

It would be grossly unfair to my colleagues in both the academy and industry to imply that no efforts have been made to remove the "black box" problem from the student's experience. But it is fair to suggest that there is very little material at the elementary level (i.e., the first two years of undergraduate study) which removes this problem in any substantial way. Thus, the following text.

In the text we will design all of the elementary circuitry for a simple digital computer. In so doing we will be able to "see" in explicit terms exactly what occurs when a program is executed. At first these programs will be written at the lowest level possible and will, therefore, be slightly tedious. We will then design both a high-level language and low-level language program to "compile" our programs written in the high-level language into low-level form.

The reader who is familiar with the above process will immediately recognize what is being suggested and might legitimately question how such a goal can be accomplished in one relatively small text. The best answer can be given by simply inviting such a reader to examine the text and see how it is done. A preliminary and partial answer can be given at this point, however. I have consistently employed what one might call a Computer Science-oriented version of Occam's Razor. William of Occam was a medieval philosopher who suggested that philosophers should seek explanations using only "necessary" reasons. Anything which could be omitted without creating a "black box" problem was, in fact, omitted. For instance, we have omitted input/output considerations on the hardware side and have settled for two-pass compiling on the software side. Such omissions and simplifications have not proven to be a problem in the sense that their omission does not leave the student mystified.

The technical background required is supplied in the first chapter. Some professors using the text may have students with background in circuit logic and design. These students could omit the first chapter with little or no risk to their understanding of the remaining chapters. However, because we have used the simplest possible logic, extensive preparation in the area of circuit logic is not necessary. The text assumes some experience with elementary programming. This experience can be prerequisite to or corequisite with the study of the text.

The material has been tested over a three-year period on students majoring in Computer Science at Graceland College in Lamoni, Iowa. The feedback from these students, all of whom are either in graduate school or industry, has been that it was the most beneficial experience of their undergraduate preparation. It is the author's hope that a similar experience will be shared by the reader.

Chapter 1

Preliminary Considerations

There are a number of terms and corresponding concepts that need to be introduced before we can present the details involved in designing a simple computer and a relatively simple compiler. Where to begin is a difficult if not impossible question and as a result, the particular order which I have chosen may appear somewhat arbitrary to the reader. However, I have attempted to order these preliminary concepts in a natural way.

First, we speak of such things as numbers, letters, characters, and so on as being "in" the computer. How is that possible? Imagine a current traveling through some appropriate material. We could associate a 0 with some specific voltage level. We will associate any deviation from 0 with a 1. When the 1 is associated with a higher voltage than the 0's voltage level, we call it positive logic and, similarly, when the 1 is associated with a lower voltage level than the 0's voltage level, we call it negative logic. This gives us a relatively easy way to simulate 0's and 1's with electricity.

What about 2,3,4,...? The apparent need for a way to represent these numbers is removed by resorting to the binary number system. In other words, instead of using a base 10 number system which requires the 10 digits 0, 1, 2, 3, 4, 5, 6, 7, 8, and 9, we use a base 2 number system which requires only 2 digits, namely, 0 and 1. As a result, the electronics we have described is capable of "representing" any number as long as we are willing to express that number in binary. Letters, A, B, C, and so on, and characters, +, ?, $, and so on, are represented by combinations of binary digits. Several standard codes have evolved for this purpose. The Extended Binary Coded Decimal Interchange Code (EBCDIC) for the letter A is 11000001. The American Standard Code for Information Interchange (ASCII) for the letter A is 01000001. For our purposes, these and other particular coding schemes are not important except as illustrations of how one might use electrical phenomena to represent numerical and linguistic phenomena.

Further reflection on the electronics would lead us to ask how one retains the various voltage levels. In other words, when the voltage on a line changes we no longer have a representation of the same number, letter, or character. And yet, we may need to retain that number, letter, or character for future reference.

Computer memory is accomplished by using an ingenious device most commonly referred to as a **flip-flop**. For the moment, imagine that a flip-flop "remembers" the last voltage level it received. This is in effect exactly what takes place. How it works will become apparent in the next chapter.

Suppose now, that we arrange several of the flip-flops in a row. For example, eight of them in a row would be capable of holding either the EBCDIC or ASCII code for the letter A. Several of these rows would constitute a computer's memory. In fact, if we were to arrange the rows in such a way so as to enable us to identify row 10 from row 11, we would have addressable memory. In Chapter 2 we will develop the circuitry for a small addressable memory.

It is also necessary for a computer to be able to access some information in memory, alter it in some way, and then return it to memory, perhaps to the same address from which it came. Where does this alteration take place? We could, of course, transfer it to some other portion of the memory, alter it there, and then transfer it back to its old location or to a new one. It is possible, however, to imagine that the portion of memory to which we transferred it for alteration contained information that we wanted to keep. For this reason, as well as many others, separate **registers** are constructed which, like memory, consist of a row of flip-flops. In Chapter 2 we will examine a computer that contains several such registers. The most important is the **accumulator**. There are others such as the **exchange register**, the **instruction register**, the **memory address register**, and the **location counter**.

In the "big" computers the memory is arranged in a manner similar to our description. However, flip-flops and other such devices have been refined to such an extent that literally thousands of them can be contained in a very small space. Modern desk-top microprocessors with semiconductor memories can hold tremendous amounts of information in the form of voltage levels (1's and 0's) which only a few years ago would have required a great deal of space in a temperature-controlled room. In fact, even the early computers were capable of storing massive amounts of information in a comparatively small space and, as a result, computer scientists and data processing specialists developed a variety of abbreviated methods for referring to computer-stored information. A binary digit is referred to as a **bit**. Eight bits treated as a group are called a **byte**. A combination of bytes, usually four, treated as a group is called a **word**. When computer people speak of a machine with a 32-bit word they are merely referring to the fact that the "rows" of flip-flops in the computer's memory are 32 flip-flops long.

Counting the number of bytes in the memory is most easily accomplished by counting in units of some power of 2. $2^{10} = 1024$ is the most common unit. A computer consisting of 1024 bytes of memory is said to have a memory capacity of 1KB where KB represents 1024 bytes. Computers are usually constructed with memory capacities measured in KB units. The number of these units is also, in most cases, a power of 2. In other words, it is common to hear of computers whose memory is described as 4KB, 8KB, 16KB, 32KB, and so on.

In addition to the memory capability, a computer must be constructed so as to allow certain operations to take place. For example, we may wish to store the information contained in the accumulator in some particular memory location. Thus a store operation must be provided. Such operations are designed in the hardware and are referred to as the **instruction set**. The computer that will be designed in the next chapter contains eight distinct instructions. Many computers have instruction sets numbering well over 100.

One instruction common to every computer is the ADD instruction. We can use it to lead into a discussion of **overflow**. When we add numbers together with pencil and paper we rarely incur an overflow problem. In a computer, however, the registers and memory words are of finite length. Therefore, repeated additions could produce a sum that would exceed the capacity of the registers. This is called overflow and is usually called to the attention of the computer user by some sort of error designation. Other instructions or sequences of instructions may also cause an overflow condition, but essentially they all represent the same phenomena, namely, the capacity of the registers has been exceeded.

Earlier we referred to addressable memory. Here we are concerned with a counting process. Such a process is accomplished electronically by circuitry, primarily flip-flops, arranged so as to produce combinations of pulses, that is, voltage levels which can be decoded into bit combinations. These bit combinations are capable of being interpreted by a **decoder**, and the results of the decoder can then be sequenced in the same fashion that numbers can be sequenced, that is, ordered. The electronics for doing this are extensive and somewhat complicated. We will make an effort to describe two simple pairs of counters and decoders in the next chapter. One pair will be used for the **location counter** and the other will be used for the **timing counter**.

We have already pointed out that the location counter is one of the registers in the computer. The timing counter on the other hand, is not a register as such, but rather, a device by which all of the operations of the computer can be sequenced. It requires a "clock." Computer clocks are devices that emit a steady stream of pulses which can be decoded into groups. The pulses in these groups are usually labeled $t_0, t_1, t_2, \ldots$ and will total to some power of 2. The power of 2 is the length of the timing cycle for the machine. For example, in the computer that we will soon be designing, the **timing counter decoder** breaks down the pulses from the clock into groups of 8, that is, 2^3. These pulses are labeled t_0, t_1, t_2, t_3, t_4, t_5, t_6, and t_7. The various instructions that the computer is to perform (e.g., the addition instruction) must not require more operations than the timing cycle will allow. The shorter the timing cycle, the faster the computer will perform a sequence of instructions.

There is one final topic that should be introduced before we begin the design process, namely, Boolean algebra. We will not attempt a complete discussion of Boolean algebra, but we will present enough preliminary information so as to allow the reader to follow the circuit design in the next chapter.

Boolean algebra, named after the logician George Boole, consists of variables whose values can only be 0 and 1 or "false" and "true" if you are interested in linguistic logic. It uses essentially four operations: the OR operation symbolized by "+"; the AND operation symbolized by ·, usually omitted or understood; the NOT operation symbolized by the overscore; and the EQUAL operation symbolized by "=". If one wishes to express the sentence *A* AND NOT *B* OR *C*, we can do so as follows:

$$A\bar{B}+C$$

The following tables show the results of these operations for various values of two inputs:

A	B	AB
0	0	0
0	1	0
1	1	1
1	0	0

A	B	$A+B$
0	0	0
0	1	1
1	1	1
1	0	1

These tables merely reflect the fact that the AND operation produces a 1 only when both inputs are 1 and the OR operation produces a 0 only when both inputs are 0. The reader will recognize a direct relationship to the use of OR and AND in everyday language. A sentence consisting of a number of clauses joined together by the conjunction AND is true only when all the clauses are true. And similarly, a sentence consisting of a number of clauses joined together by the disjunction OR is false only when all the clauses are false.

The number of combinations of input values for N variables is equal to 2^N. Suppose we had 3 variables, A, B, and C. These combinations could be listed as follows:

A	B	C
0	0	0
0	0	1
0	1	1
0	1	0
1	1	0
1	1	1
1	0	1
1	0	0

Let us suppose that a Boolean expression is desired which will yield a 1 every time the inputs sum numerically to 2. Because we do not know such an expression let us designate it by "f." We can now set up the following table:

A	B	C	f
0	0	0	0
0	0	1	0
0	1	1	1
0	1	0	0
1	1	0	1
1	1	1	0
1	0	1	1
1	0	0	0

There is a relatively simple technique for deriving the expression f. This technique, known as the **sum of products** approach, requires that we express f as a number of ANDed expressions which are all ORed together. Now, the only way the ANDed expressions will produce a 1 is if all the inputs are 1. Looking at the table we see that when $A=0$, $B=1$, and $C=1$, f should be a 1. Therefore we will negate the A and AND it with B and C. We obtain:

$$\bar{A}BC$$

Two other ANDed expressions are possible, namely,

$$AB\bar{C}$$

and

$$A\bar{B}C$$

The expression for f is now obtained by ORing these ANDed expressions together; that is,

$$f=\bar{A}BC+AB\bar{C}+A\bar{B}C$$

The reader may easily verify that this expression does indeed produce a 1 only when the numerical sum of A, B, and C is exactly 2.

Methods are available that may allow simplification of expressions such as the one we have obtained for f. We will concern ourselves only with methods that involve the application of some useful theorems from Boolean algebra. The theorems are as follows:

(1) $A+AB=A$
(2) $AB+A\bar{B}=A$
(3) $AB+\bar{A}C=AB+\bar{A}C+BC$

A simplified proof for these theorems can be easily accomplished by listing the individual variables and all the combinations of input values. The left and right sides of each theorem can then be evaluated for each combination of the input values. If they produce the same result for each combination of input values, the theorem is established. The proof for theorem (3) is given as follows:

A	B	C	$AB+\bar{A}C$	$AB+\bar{A}C+BC$
0	0	0	0	0
0	0	1	1	1
0	1	1	1	1
0	1	0	0	0
1	1	0	1	1
1	1	1	1	1
1	0	1	0	0
1	0	0	0	0

There are, of course, many other theorems in Boolean algebra but these will be sufficient for our purposes. Their application in reducing expressions similar to f above will become apparent in the next chapter.

Our present task is to design a computer, a high-level language for that computer, and finally, a compiler for the programs written in the high-level language. Almost every why-question that a student of Computer Science might ask should be answered in some degree once the task is complete. Therefore, our why-computer will be referred to as the WHYCO and the programming language that we will eventually design will be referred to as PL/W. Let us begin.

Chapter 2

Hardware Design For a Simple Digital Computer

2-1 Instructions

A digital computer is a machine capable of performing a set of conceptually simple instructions, for example, addition. Relatively complex problems can, however, be solved by repeated use of just a few basic instructions in a program. With this in mind, we will select the following set of basic instructions which will be incorporated into the WHYCO computer.

ADDITION

SUBTRACTION

SHIFT LEFT ONE

TRANSFER

TRANSFER ON NEGATIVE

STORE ACCUMULATOR

CLEAR ACCUMULATOR

STOP

Eventually we will define the above instructions in a symbolic form. Meanwhile, however, it will be to our advantage to describe each of them in the simplest language possible. As you read these descriptions, assume that the word **register** refers to something similar to the visual display on a pocket calculator. It is not quite that simple but this analogy will provide a sufficient conceptual basis for describing these instructions at this stage of the design process.

DEFINITION ADDITION (ADD) This instruction causes the contents of the memory location specified by the memory address register to go into exchange register. The contents of the exchange register and the accumulator are then transferred to the adder and this sum is placed in the accumulator.

Before defining the subtraction instruction, we must discuss "signed two complement arithmetic" (S2C). Let us clarify at the beginning that this is not intended as a generalized treatment of S2C arithmetic. It is introductory in nature and slanted toward the WHYCO.

Why is S2C arithmetic useful? To answer this question in complete detail would carry us beyond our current concern. However, a brief justification for its use is necessary because eventually we will design for our computer a unit called the **arithmetic unit**. All arithmetic operations will be performed in this unit. To make our overall design task more manageable, we included only two arithmetic operations, ADDition and SUBtraction. S2C arithmetic eliminates the need, in the arithmetic unit, for both an adder and a subtractor.

Another reason for using S2C arithmetic involves the detection of positive and negative numbers. In S2C arithmetic, no sensing of the sign is required, because the sign is added together with the rest of the register.

There are several other methods for representing negative numbers. Two of them are "signed magnitude" and "signed one's complement." Positive numbers are identical in all of these methods. These two methods together with S2C are by no means exhaustive but cover at least 95 percent of existing fixed word length machines.

Before supplying a mathematical proof for S2C arithmetic, let us illustrate its use with a few simple examples. As we will soon see, the WHYCO will have a "word" length of 9 binary digits (bits), and therefore the following examples use 9-bit integer numbers.

Consider a number with the leftmost bit of the register as the sign bit, 0 for positive and 1 for negative. For example:

$$A = (011000000)_2 = +(2^7 + 2^6) = 192_{10}$$

To represent $-A$ we must do more than simply change the sign bit. Changing only the sign bit would give the value of $-A$ in signed magnitude form. The method, or algorithm, is to complement each bit, including sign, and add 1 into the least significant bit position. We must, of course, allow carrys to propagate up the number as far as required. For our case,

$$\begin{aligned} -A &= (1001111111)_2 + (000000001)_2 \\ &= (101000000)_2 \end{aligned}$$

Let us consider two examples of addition.

EXAMPLE 1:

$$\begin{array}{rr} A = & 011000000 \\ +(-A) = & \underline{101000000} \\ (1) & 000000000 \end{array}$$

Observe that a carry propagated right "through" the sign bit. Ignore it!

EXAMPLE 2: The second case is:

$$\begin{array}{rr} A = & 011000000 \\ +A = & \underline{011000000} \\ & 110000000 \end{array}$$

The sum of two positive numbers is never negative, therefore, the sum of two positive numbers should not carry into the sign position. Therefore, this example represents an overflow error.

Let us now consider two examples of subtraction using:

$$A = 000000111$$
$$B = 000000101$$

EXAMPLE 3: For A subtract B we have:

$$\begin{array}{r} 000000111 \\ \\ \underline{111111011} \\ 000000010 \end{array} \qquad 7-5=2$$

EXAMPLE 4: For B subtract A we have:

$$\begin{array}{r} 000000101 \\ \\ \underline{111111001} \\ 111111110 \end{array} \qquad 5-7=-2$$

Note that -2 is 111111110.

Consider now a binary integer written in symbolic form. We will let B_i represent the ith bit position. For an N-bit binary number we have the following decimal equivalent:

$$B_n*2^n + B_{n-1}*2^{n-1} + \cdots + B_2*2^2 + B_1*2^1 + B_0*2^0$$

If we now subtract a second number whose decimal value is represented as:

$$A_n*2^n + A_{n-1}*2^{n-1} + \cdots + A_2*2^2 + A_1*2^1 + A_0*2^0,$$

we obtain:

$$(B_n - A_n)*2^n + (B_{n-1} - A_{n-1})*2^{n-1} + \cdots$$
$$+ (B_2 - A_2)*2^2 + (B_1 - A_1)*2^1 + (B_0 - A_0)*2^0 \quad \textbf{(1)}$$

Let us now perform the same operation using the S2C algorithm. First we complement the number to be subtracted. Its decimal value can now be represented as follows:

$$\bar{A}_n*2^n + \bar{A}_{n-1}*2^{n-1} + \cdots + \bar{A}_2*2^2 + \bar{A}_1*2^1 + \bar{A}_0*2^0$$

The second step in the algorithm calls for the addition of a one (1) in the least significant bit position. The decimal value that results is expressed as follows:

$$\bar{A}_n*2^n + \bar{A}_{n-1}*2^{n-1} + \cdots + \bar{A}_2*2^2 + \bar{A}_1*2^1 + \left(\bar{A}_0 + 1\right)*2^0$$

The final step is simply the addition of our last result to the other number. The decimal value is expressed as follows:

$$\left(B_n + \bar{A}_n\right)*2^n + \left(B_{n-1} + \bar{A}_{n-1}\right)*2^n + \cdots + \quad \textbf{(2)}$$
$$\left(B_2 + \bar{A}_2\right)*2^2 + \left(B_1 + \bar{A}_1\right)*2^1 + \left(B_0 + \bar{A}_0 + 1\right)*2^0$$

By comparing equation (2) with (1) and eliminating the B_i coefficient, it is clear that the following equation must hold if the two are equal:

$$-A_n*2^n - A_{n-1}*2^{n-1} + \cdots + -A_2*2^2 - A_1*2^1 - A_0*2^0$$
$$= \bar{A}_n*2^n + \bar{A}_{n-1}*2^{n-1} + \cdots + \bar{A}_2*2^2 + \bar{A}_1*2^1 + \left(\bar{A}_0 + 1\right)*2^0$$

This can be rewritten as:

$$-\left(A_n+\bar{A}_n\right)*2^n-\left(A_{n-1}+\bar{A}_{n-1}\right)*2^{n-1}-\cdots$$
$$-\left(A_2+\bar{A}_2\right)*2^2-\left(A_1+\bar{A}_1\right)*2^1-\left(A_0+\bar{A}_0+1\right)*2^0=0 \qquad (3)$$

We now make use of the following:

$$A_i+\bar{A}_i=1$$
$$A_0+\bar{A}_0+1=2$$

A binary digit cannot be a "2." Therefore, it will be treated as a "0" with a "1" carry into the next bit position. For any bit position, say the ith position, we will then have:

$$A_i+\bar{A}_i+1=2,$$

which will once again be treated as a "0" with a "1" carry into the next bit position. This will continue until the nth bit position is "0" and "1" carry is propagated to the $n+1$th bit position, which does not exist, that is, is off the register. Therefore, all the coefficients of the powers of 2 in (3) are zero, as they should be. We are now prepared to define the subtraction instruction as well as the rest of the WHYCO instruction set.

DEFINITION SUBTRACTION (SUB) The contents of the memory location specified by the memory address register go into the exchange register. The contents of the exchange register are complemented. Then, the contents of the exchange register and the accumulator are transferred to the adder and the sum is placed in the accumulator. A 1 with a binal point of 0 is placed in the exchange register. Once again the contents of the exchange register and the accumulator are transferred to the adder and the sum is placed in the accumulator.

DEFINITION SHIFT LEFT ONE (SLO) The contents of the accumulator are transferred to the exchange register. Then the contents of the accumulator and the exchange register are transferred to the adder. The overflow carry bit is ignored. The sum is the original binary sequence shifted left one bit and is in the accumulator.

DEFINITION TRANSFER (TRA) Copy the contents of the memory address register into the location counter.

DEFINITION TRANSFER ON NEGATIVE (TRN) If the accumulator contains a negative number, copy the contents of the memory address register into the location counter; otherwise, do nothing.

DEFINITION STORE ACCUMULATOR (STA) Copy the contents of the accumulator into the exchange register. Then copy the contents of the exchange register into the memory location specified by the memory address register.

DEFINITION CLEAR ACCUMULATOR (CLA) Clear the accumulator to zeros.

DEFINITION STOP (STP) Reset the STOP/RUN flip-flop to zero. Flip-Flops will be discussed in Section 2-6.

2-2 The Fetch-Execute Cycle

We have just spelled out a rather detailed word description of the manner in which the WHYCO will execute each of the eight selected instructions. Which instruction will be executed at any instant in time depends on which one happens to be in the **Instruction Register**. How do we then manage to get the proper instructions into and out of the instruction register?

Instructions are usually stored sequentially in memory. The location of the next sequential instruction to be executed in WHYCO is found in the location counter.

Thus, after the execution of a particular instruction has been completed, WHYCO must go through a fetch cycle to fetch the next instruction from memory and copy it into the instruction register. Note that any time we copy from memory or from a register, the contents are preserved.

Thus, the order of operation in WHYCO will be

1. FETCH first instruction
2. EXECUTE first instruction
3. FETCH second instruction
4. EXECUTE second instruction
5. and so on

We must, therefore, describe the FETCH cycle.

DEFINITION FETCH CYCLE The contents of the location counter are first copied into the memory address register. Next, the contents of the memory location specified by the memory address register are copied into the exchange register. The instruction portion of the contents of the exchange register is copied into the instruction register, and the address portion of the contents of the exchange register is copied into the memory address register. Finally, a positive one is added to the present contents of the location counter, so it will contain the location of the subsequent instruction to be fetched from memory.

2-3 Subcommand Generation

Thus far, we have discussed the role that each of the registers in the WHYCO plays by discussing in some detail the orderly manner in which information must be moved around and operated upon during the FETCH and EXECUTE cycles. Because the computer should be capable of operating without human intervention, some type of master control is needed to assure that

1. Only those operations or subcommands necessary to completely execute a particular instruction in the instruction register, or fetch another instruction from memory, are performed.
2. These operations or subcommands occur in the proper sequence.

This control is the function of the **Subcommand Generator**. With the aid of a clock and a timing counter to count clock pulses we will be able to decode the timing information. Likewise, the instruction register can also be decoded. With knowledge of these two facts (time and instruction), we should be in a position to combine them and form various sequences of subcommands.

2-4 Symbolic Design

Before we attempt a more detailed design of WHYCO, it will be helpful to restate our description of the instructions and fetch cycle in symbolic form. We will simply rewrite our word descriptions using these symbols.

Our symbolic design symbols are defined below.

(X)	Parentheses denote contents ot a register, for example, Register X.
OP(X)	OP denotes the operand part of register X.
AD(X)	AD denotes the address part of register X.
SB(X)	SB denotes the sign bit part of register X.

M ⟨MAR⟩ This symbol denotes the *location* of a memory word addressed by the memory address register.

$(A) \Rightarrow B$ Double-line arrows denote a transfer of bits from one register to another—for example, A to B.

$V: (A) \Rightarrow B$ A colon following a variable denotes the occurrence of the subsequent statement when the value of the variable is a 1.

$\overline{X}$ The bar denotes the complementing of the bit string X.

We are now ready to symbolically define the sequence of subcommands necessary to execute each of the instructions in WHYCO. The reader should carefully compare these symbolic definitions with the verbal definition given above. Eventually the symbolic definition will be sufficient to indicate any operation and reference to the verbal descriptions will no longer be necessary.

000 ADD	$(M<MAR>) \Rightarrow XR$
	$(XR)+(AC) \Rightarrow AC$
001 SUB	$(M<MAR>) \Rightarrow XR$
	$(\overline{XR}) \Rightarrow XR$
	$(XR)+(AC) \Rightarrow AC$
	$1 \Rightarrow XR$
	$(XR)+(AC) \Rightarrow AC$
010 SLO	$(AC) \Rightarrow XR$
	$(XR)+(AC) \Rightarrow AC$
011 TRA	$(MAR) \Rightarrow LC$
100 TRN	$AC_1:(MAR) \Rightarrow LC$
	$(AC) \Rightarrow XR$
101 STA	$(XR) \Rightarrow M<MAR>$
110 CLA	$0 \Rightarrow AC$
111 STP	$0 \Rightarrow STOP/RUN\ FF$
FETCH CYCLE	$(LC) \Rightarrow MAR$
	$(M<MAR>) \Rightarrow XR$
	$OP(XR) \Rightarrow IR$
	$AD(XR) \Rightarrow MAR$
	$(LC)+1 \Rightarrow LC$

Because some of the subcommands are used more than once, and in more than one instruction, it will be helpful to make up a list of the necessary subcommands. This list will be used when we turn to the design of the subcommand generator.

1. (M<MAR>) ⇒XR
2. (XR)+(AC) ⇒AC
3. $(\overline{\text{XR}})$ ⇒XR
4. 1 ⇒XR
5. (AC) ⇒XR
6. (MAR) ⇒LC
7. AC_1: (MAR) ⇒LC
8. (XR) ⇒M(MAR)
9. 0 ⇒AC
10. 0 ⇒STOP/RUN FF
11. (LC) ⇒MAR
12. OP(XR) ⇒IR
13. AD(XR) ⇒MAR
14. (LC)+1 ⇒LC

2-5 Logic

In this section we will discuss as briefly as possible the actual circuit mechanisms that will be used in the WHYCO. These will not be sophisticated mechanisms, but they will be sufficient for our purposes. Actually the sophistication referred to is achieved by combining the basic elements in different ways. For example, it is customary to design circuits with NANDs or NORs which are simply combinations of ANDs and INVERTERs or ORs and INVERTERs, respectively. We will be content to use the basic gating elements, that is, AND, OR and the INVERTER.

The AND gate simply combines the inputs in such a way so as to produce a 1 output only when all the inputs are 1. In this context 1 refers to a voltage level that is either high (positive logic) or low (negative logic) when compared with the normal voltage (0) on the line. The Boolean operator for the AND gate is the dot (·). Therefore $X_1 \cdot X_2 \cdot \cdots \cdot X_n$ will represent the ANDing of n input signals. The gating symbol is as follows:

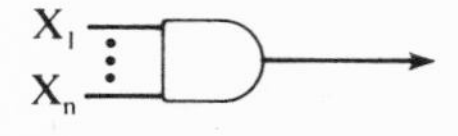

The OR gate simply combines the inputs in such a way so as to produce a 1 output when at least one of the inputs is a 1. The Boolean operator for the OR gate is the plus (+). Therefore $X_1 + X_2 + \cdots + X_n$ will represent the ORing of n input signals. The gating symbol is as follows:

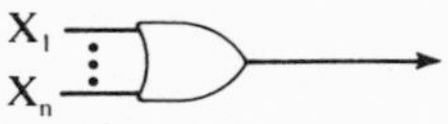

The INVERTER changes a 1 to 0 or a 0 to 1. The Boolean operator for the INVERTER is a bar (-) written over the variable name. Therefore, $\overline{X_1}$ will represent the inverting of the X_1 signal. The gating symbol is as follows:

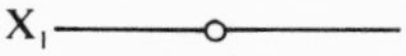

2-6 Elementary Flip-Flop Design

The component most essential to our design problem is the **flip-flop**. This is an ingenious device consisting of the elements described in Section 1.5 whose function is to "remember" a logic value which partially depends upon an input. We will attempt to construct an extremely simple version of such a device in this section. Since there are many different kinds of flip-flops, we will simply describe the more complex type used in the WHYCO.

Let us suppose that it is possible to construct a device which will "retain" a "pulse" (1) when "set" and will "retain" a "no pulse" (0) when "reset." The circuit for such a device must be sequential due to the fact that combinational circuits simply process the input pulses into an output pulse. However, we must retain the input pulse in the circuit. We will suggest, therefore, that the output is "recycled" as input. In other words, we have a feedback loop.

Let us assume that the value of the device (hereafter referred to as a SET-RESET flip-flop) at time T is indicated by $F(T)_{in}$ and the value of the device some time later is indicated by $F(T)_{out}$. We have the following table:

S	R	$F(T)_{in}$	$F(T)_{out}$
0	0	0	0
0	0	1	1
0	1	1	1
0	1	0	0
1	1	0	Ø
1	1	1	Ø
1	0	1	1
1	0	0	1

Using the **sum of products approach** we can write the following:

$$F(T)_{out} = \overline{S}\,\overline{R}F(T)_{in} + S\overline{R}F(T)_{in} + S\overline{R}\,\overline{F}(T)_{in}$$

Using the fact that $AB+A\overline{B}=A$, we can let $A=\mathrm{S}\overline{\mathrm{R}}$ and $B=\mathrm{F(T)}$ to obtain:

$$\mathrm{F(T)_{out}}=\overline{\mathrm{S}}\overline{\mathrm{R}}\mathrm{F(T)_{in}}+\mathrm{S}\overline{\mathrm{R}}.$$

A further simplification can be obtained by using the fact that $AB+\overline{A}C=AB+\overline{A}C+BC$. We can let $A=\mathrm{S}$, $B=\overline{\mathrm{R}}$, and $C=\overline{\mathrm{R}}\mathrm{F}$ to obtain:

$$\mathrm{F(T)_{out}}=\overline{\mathrm{S}}\overline{\mathrm{R}}\mathrm{F(T)_{in}}+\mathrm{S}\overline{\mathrm{R}}+\overline{\mathrm{R}}\mathrm{F(T)_{in}}.$$

Using the fact that $A+AB=A$ we can let $A=\overline{\mathrm{R}}\mathrm{F(T)_{in}}$ and $B=\overline{\mathrm{S}}$ to obtain

$$\mathrm{F(T)_{out}}=\mathrm{S}\overline{\mathrm{R}}+\overline{\mathrm{R}}\mathrm{F(T)_{in}}.$$

Our circuit is therefore as shown in Figure 2-1.

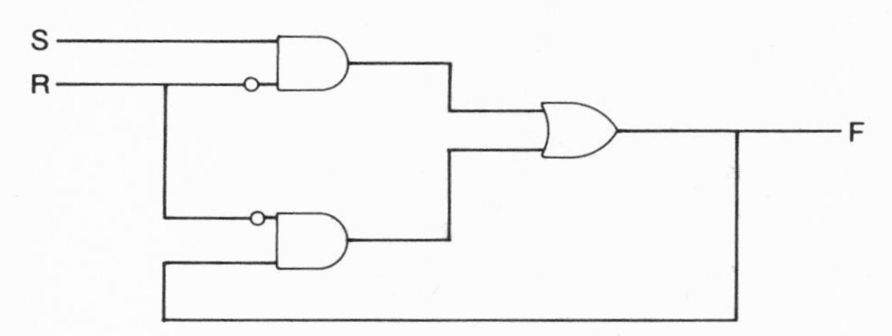

FIGURE 2-1 SET-RESET Flip-Flop

A common technique for displaying the behavior of a circuit such as the one in Figure 2-1 is to use the state table. (See Figures 2-2 and 2-3) This table relates all the combinations of the input values, the output values, and the current status of the flip-flop, that is, is it "remembering" a 1 or a 0, in such a way so as to allow one to trace through the SET-cycle or the RESET-cycle. The values along the top of the table represent the various combinations of values for SET (S) and RESET (R). The one and zero entries in the table represent the value of the flip-flop at the time of the SET or RESET pulse. If the value is circled, it represents stability. If it is not circled, the flip-flop transitions to the other state. When the flip-flop assumes a stable condition in row one it is in state S_0, and, similarly, when the flip-flop assumes a stable condition in row two, it is in state S_1.

Let us now trace the set cycle. To begin with S=0, R=0, F=0. When S=1, R=0, F=0 we can see that the upper AND-gate in Figure 2-1 produces a 1 and then the OR-gate will produce a one, so that S=1, R=0, F=1. Now, however, the set line has returned to zero, so that S=0, R=0, F=1. By tracing these values through the flip-flop, we see that the lower AND-gate will now produce a one and therefore the OR-gate will again produce a one. This will continue to happen until the RESET line is pulsed. The reader is encouraged to work though a similar trace for the reset cycle. Figures 2-2 and 2-3 are merely convenient ways to summarize the trace of these cycles.

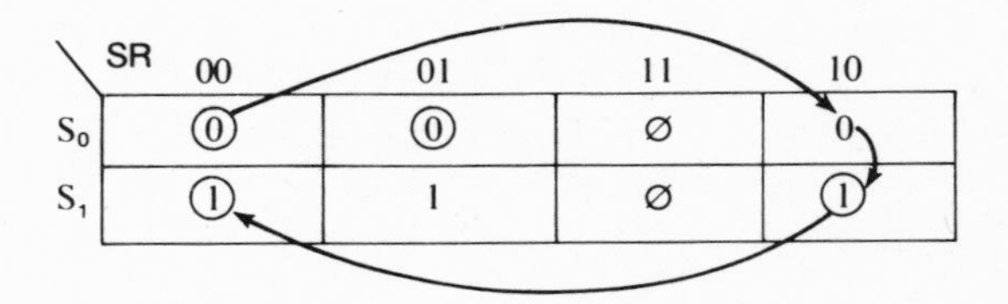

FIGURE 2-2 SET Cycle for SET-RESET Flip-Flop

SR	00	01	11	10
S_0	0	0	Ø	0
S_1	1	1	Ø	1

FIGURE 2-3 RESET Cycle for SET-RESET Flip-Flop

As we mentioned earlier, the design for the flip-flop used in the WHYCO is considerably more complex. The clocked JK flip-flop is well known and uses a clock pulse which provides a way to coordinate the timing of the flip-flops behavior with the rest of the WHYCO computer. The SET and RESET lines on this flip-flop are referred to as J and K, respectively. It has an added feature in that simultaneous pulsing of the SET (J), RESET (K), and clock line will result in complementation of the flip-flop. Assuming a clock pulse of "1" we have the following table:

J	K	$F(T)_{in}$	$F(T)_{out}$	DESCRIPTION
0	0	0	0	No effect
0	0	1	1	No effect
0	1	1	0	Reset to 0
0	1	0	0	Reset to 0
1	1	0	1	Complement
1	1	1	0	Complement
1	0	1	1	Set to 1
1	0	0	1	Set to 1

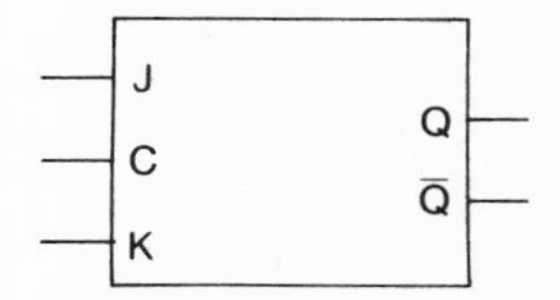

FIGURE 2-4 Symbol for the Clocked J-K Flip-Flop

The output from the clocked JK flip-flop is referred to by Q and will be provided in complemented and uncomplemented form. We will therefore symbolize the flip-flop as in Figure 2-4.

Before we proceed to the design of the functional building blocks, the size of the computer and the various registers should be specified.

Because we judiciously chose 8 instructions, three flip-flops will be needed ($2^3 = 8$) in the instruction register to hold the binary code. Note that 000 through 111 provide for eight distinct codes.

Since more than 4, but less than 8, subcommands are needed in a particular sequence, three flip-flops will be needed in the timing counter. The design of the timing counter will illustrate the need for the three flip-flops.

We will temporarily select a 64 word memory, which means that six flip-flops ($2^6 = 64$) will be needed for the memory address register and the location counter. Here again, the design of the memory address register and the location counter will illustrate the need for the six flip-flops.

Because we will store a single-address and an instruction in one word of memory, each memory word must be 9 flip-flops long, three for the instruction and six for the address.

Because data will also be stored in memory, numbers 9 bits in length can be used. Therefore, the exchange register and the accumulator must be 9 flip-flops long.

The adder must then be capable of adding two 9 bit numbers, where the left-hand bit will be designated as the sign bit. Initially, no provision will be made to signal the occurrence of an overflow bit. Therefore, the user will not know when the capacity of the WHYCO is exceeded during an addition.

2-7 The Whyco Computer

We will now proceed to the actual design of the WHYCO. We start with Figure 2-5, which shows the general design of the basic units of the WHYCO and the linkages between them. The numbers in parentheses indicate the order in which the individual elements are discussed in the text.

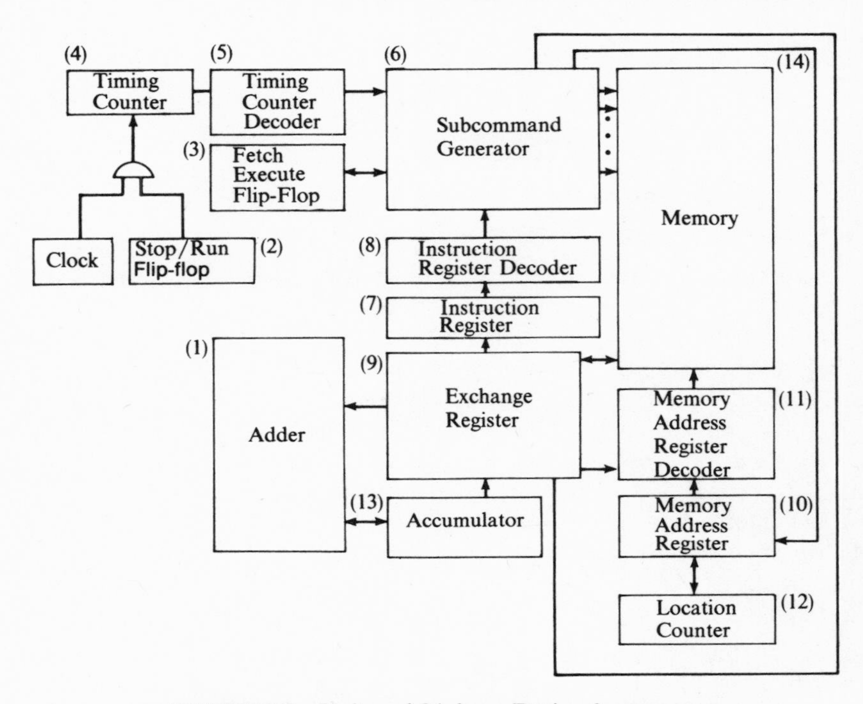

FIGURE 2-5 Unit and Linkage Design for WHYCO

As has been suggested, the WHYCO will need only an adder. This is because of its use of S2C arithmetic. Therefore, our arithmetic units will consist of the circuits necessary to complete a simple addition of two registers, namely, the accumulator (AC) and the exchange register (XR). For the ith bit position, and letting S_i and C_i stand for the sum and carry bits, respectively, we would have the following:

C_{i-1}	AC_i	XR_i	S_i	C_i
0	0	0	0	0
0	0	1	1	0
0	1	1	0	1
0	1	0	1	0
1	1	0	0	1
1	1	1	1	1
1	0	1	0	1
1	0	0	1	0

Using the sum of products approach, we can write the following Boolean equations:

$$
\begin{aligned}
S_i &= \bar{C}_{i-1}\cdot \overline{AC}_i\cdot XR_i + \bar{C}_{i-1}\cdot AC_i\cdot \overline{XR}_i \\
&\quad + C_{i-1}\cdot AC_i\cdot XR_i + C_{i-1}\cdot \overline{AC}_i\cdot \overline{XR}_i \\
C_i &= \bar{C}_{i-1}\cdot AC_i\cdot XR_i + C_{i-1}\cdot AC_i\cdot \overline{XR}_i \\
&\quad + C_{i-1}\cdot AC_i\cdot XR_i + C_{i-1}\cdot \overline{AC}_i\cdot XR_i \\
&= \bar{C}_{i-1}\cdot AC_i\cdot XR_i + C_{i-1}\cdot AC_i \\
&\quad + C_{i-1}\cdot \overline{AC}_i\cdot XR_i
\end{aligned}
$$

The implementation for these equations is given in Figures 2-6 and 2-7 using simple combinational circuits.

Because of space and simplicity we will use a diagram showing only the input and output to the above circuitry for each bit position (see Figure 2-8).

This device is a **full adder** and can be linked together to form the **arithmetic unit** (see Figure 2-9). Two problems arise at this point. First, where does C_0 come from? Actually the value of C_0 does not exsist. This difficulty can be remedied by replacing the **full adder** in position one with a **half adder**. The half adder assumes no carry-in but does produce a

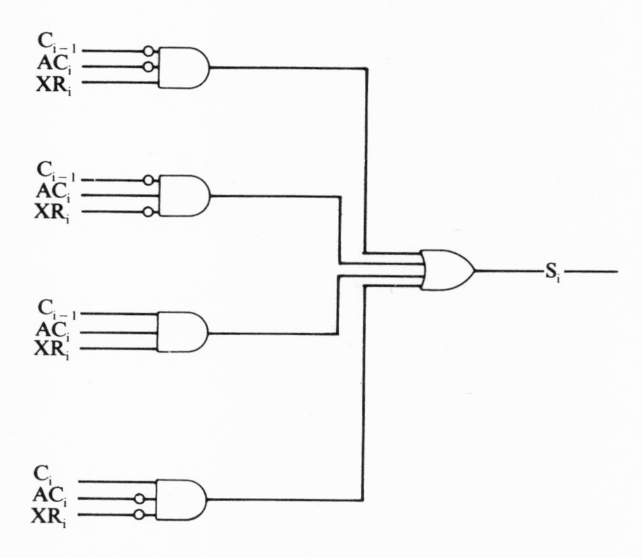

FIGURE 2-6 Sum-Bit Circuitry for a Full Adder

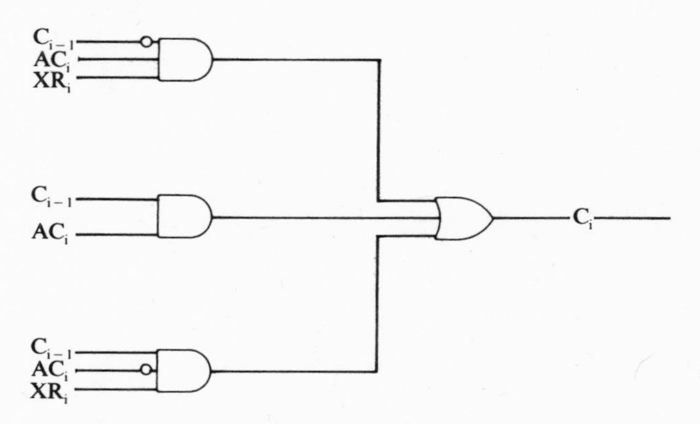

FIGURE 2-7 Carry-Bit Circuitry for a Full Adder

carry-out along with the sum bit. We have:

AC_i	XR_i	S_i	C_i
0	0	0	0
0	1	1	0
1	1	0	1
1	0	1	0

and the following Boolean equations:

$$S_i = \overline{AC}_i \cdot XR_i + AC_i \cdot \overline{XR}_i$$
$$C_i = AC_i \cdot XR_i$$

The circuit design is shown in Figures 2-10 and 2-11.

Secondly, what do we do with C_9? Actually this indicates an overflow when it is equal to 1. As was indicated in Section 2.6, we will ignore it, but we should point out that it is ignored only to keep our design problem simple. In a more complex machine it would be used to signal an execution error to the user.

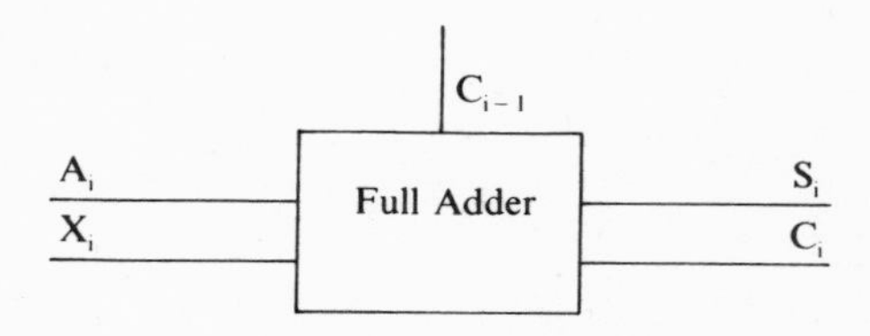

FIGURE 2-8 Symbol for a Full Adder

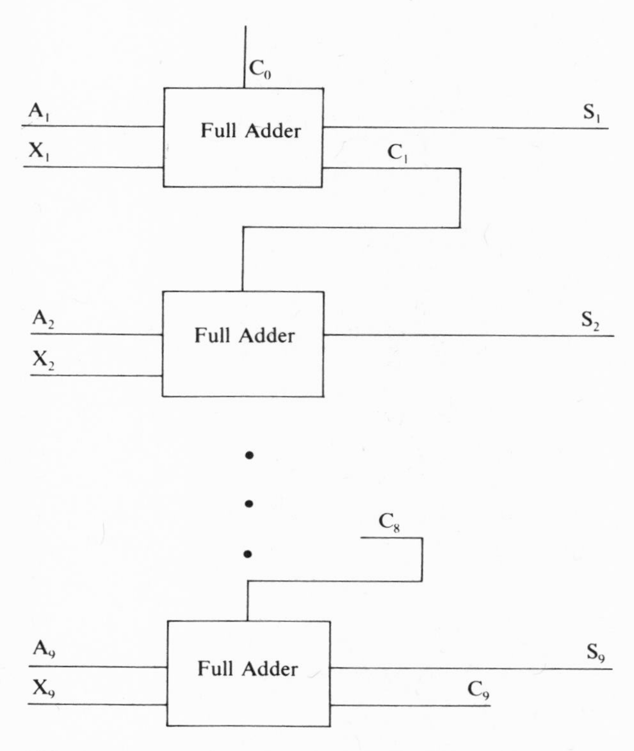

FIGURE 2-9 Arithmetic Unit Design using Full Adders

Our final design configuration for the arithmetic unit is shown in Figure 2-12.

Adders have been designed which are much more efficient than the one represented in Figure 2-12. The interested reader can find these designs in any standard text on computer architecture.

The **stop-run** flip-flop is simply the device by which we turn the computer on and off. It must be connected to the clock pulse so that when it is on it will transmit the clock pulse to the timing counter (Figure 2-13).

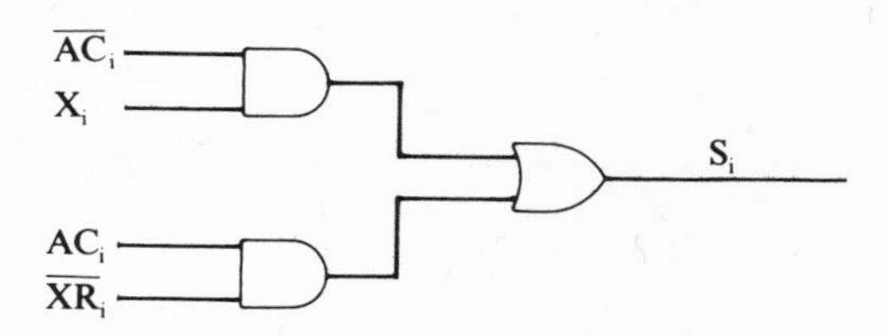

FIGURE 2-10 Sum-Bit Circuitry for a Half Adder

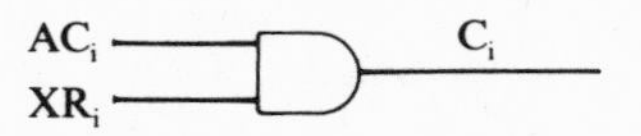

FIGURE 2-11 Carry-Bit Circuitry for a Half Adder

The last timing pulse in the **sub-command generator** will trigger the **fetch-execute** flip-flop in such a way so as to alternately fetch the next instruction and then execute it. The last timing pulse is t_7 (see Figure 2-16) and therefore this must be connected to the **fetch-execute** flip-flop as shown in Figure 2-14.

Figure 2-15 displays the arrangement of flip-flops which will produce 3 binary outputs for a combination of 2^3 or 8 separate timing pulses.

The JK flip-flop responds to simultaneous pulses on both the J and K lines by complementing itself. Assuming this takes place when the clock pulse falls, and noting that the clock pulse for the second and third flip-flop is actually the output of the first and second flip-flop, respectively, we have the timing diagram shown in Figure 2-16.

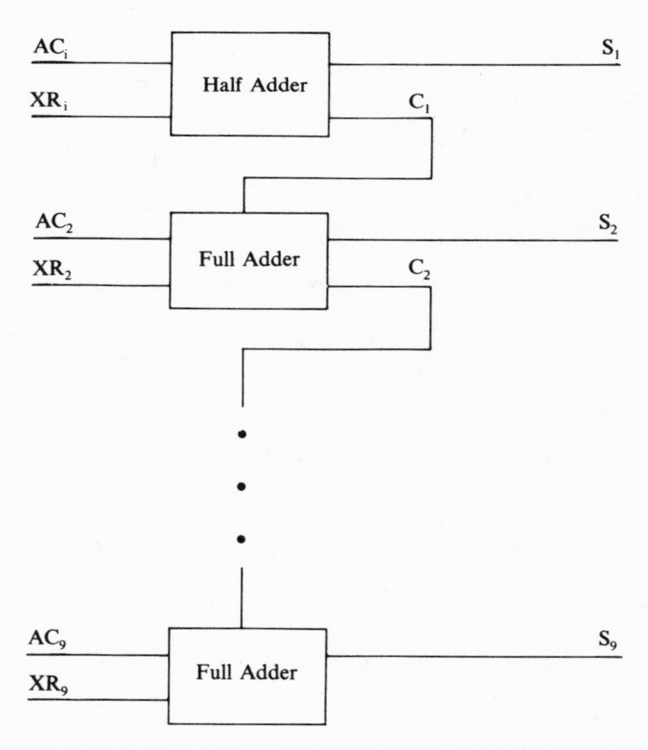

FIGURE 2-12 Arithmetic Unit Design Using Half and Full Adders

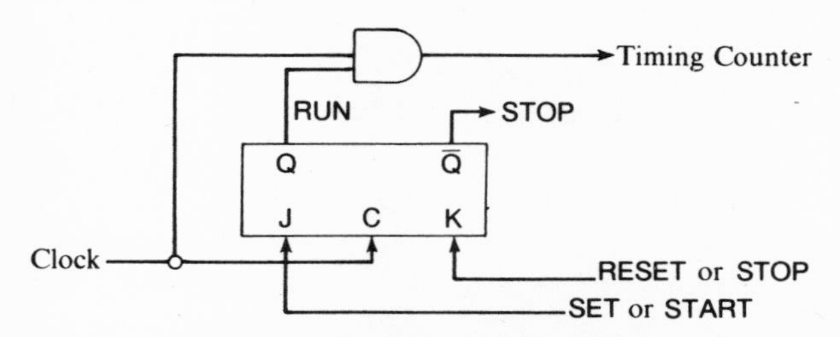

FIGURE 2-13 STOP-RUN Flip-Flop

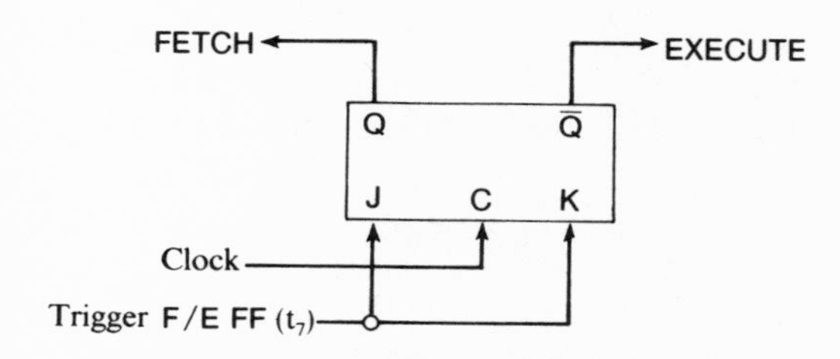

FIGURE 2-14 FETCH EXECUTE Flip-Flop

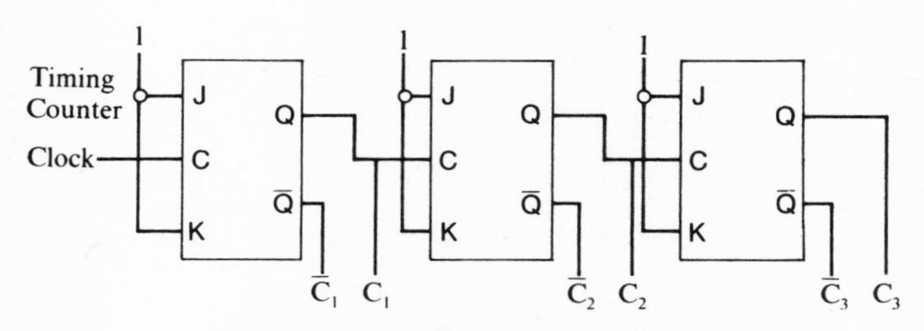

FIGURE 2-15 3-Bit Timing Counter

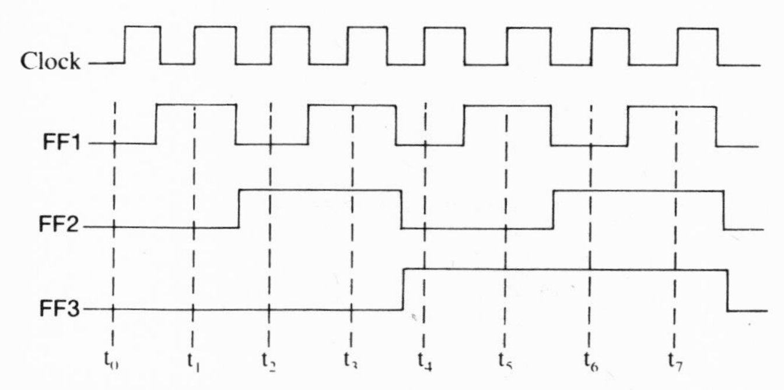

FIGURE 2-16 Timing Diagram for 3-Bit Timing Counter

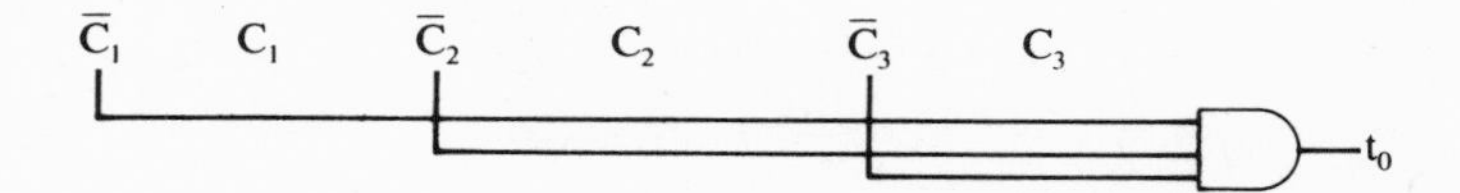

FIGURE 2-17 Decoder Circuitry for the First Timing Pulse

Capturing the eight distinct combinations of pulse values from the **timing counter**, indicated in Figure 2-16 by t_i, $i=0,\ldots,7$, is achieved by the **timing counter decoder**. The timing counter decoder is simplified diagrammatically by the use of single lines. For example, the complete diagram for t_0 would be as in Figure 2-17. It is common, however, to use the simplified single line diagram, and this procedure will be followed throughout this text (see Figure 2-18).

The subcommand generator uses the results of three other components of the WHYCO. They are: (1) fetch-execute flip-flop, (2) timing counter decoder, and (3) instruction register decoder. We have already completed the design of (1) and (2), and could at this point turn to the design of (3). However, the instruction register decoder depends on the instruction register which in turn depends on the subcommand generator. It is optional, therefore, as to which we do first, subcommand generator or instruction register decoder. We have elected to do the subcommand generator.

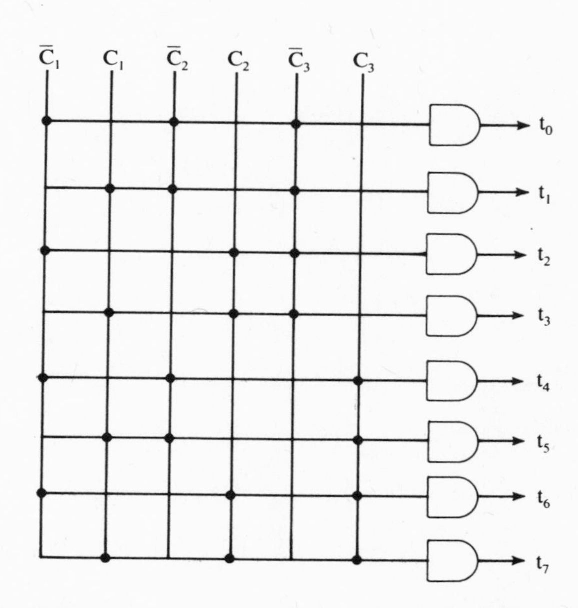

FIGURE 2-18 Timing Counter Decoder for 3-Bit Timing Counter

This device simply sequences the appropriate subcommands so as to insure that the current instruction will in fact be executed. For example, if the instruction register decoder produces a one on the SUB line (it is important that the reader realizes that a one on SUB also means there is a zero on ADD, SLO, TRA, TRN, STA, CLA, and STP), and if at the same time there is a one on the EXECUTE line, then the first five timing pulses, t_0, t_1, t_2, t_3, t_4, will cause the appropriate set of subcommands to be executed. From this point on in our design we will see the various subcommand lines, which are at the left of Figure 2-19, appearing in the various registers.

The **instruction register** always receives its content from the first three bit positions of the exchange register. Therefore, we need to connect the output lines from the first three bit positions of the exchange register to the *J* and *K* terminals of the instruction register. We do not, however, want the instruction register to accept these values unless the line OP(XR)⇒IR which comes from the subcommand generator has a one on it. An easy way to accomplish this is to use the OP(XR)⇒IR line for the clock pulse. The *JK* flip-flop is designed to ignore input on its set and reset terminals unless the clock pulse is a one. Figure 2-20 shows the circuit design for the instruction register.

The **instruction register decoder** (see Figure 2-21) works on the same principle as the timing counter decoder. This does not mean, however, that the timing counter and the instruction register do the same thing. In the timing counter the flip-flops are arranged so that the first flip-flop "clocks" the second, and so on, whereas in the instruction register the flip-flops are "clocked" by the corresponding flip-flop in the exchange register. As was the case with the timing counter decoder, we have used the single line diagram. (Compare Figures 2-17 and 2-18.)

The **exchange register** lives up to its name in the fullest sense of the word. In nearly all cases, when information is transferred from one location to another, it passes through the exchange register. For example, a word from a program will consist of an operation and an operand address. Before the operation is transferred to the instruction register and the operand address is transferred to the memory address register, the entire word is placed in the exchange register by the second subcommand of the FETCH cycle.

Perhaps the best way to approach the design of the exchange register is to locate on the subcommand generator (Figure 2-19) every output line of the form ______⇒XR. They are as follows:

1. $\overline{XR}$⇒XR
2. 1⇒XR
3. (M <MAR>)⇒XR
4. (AC)⇒XR

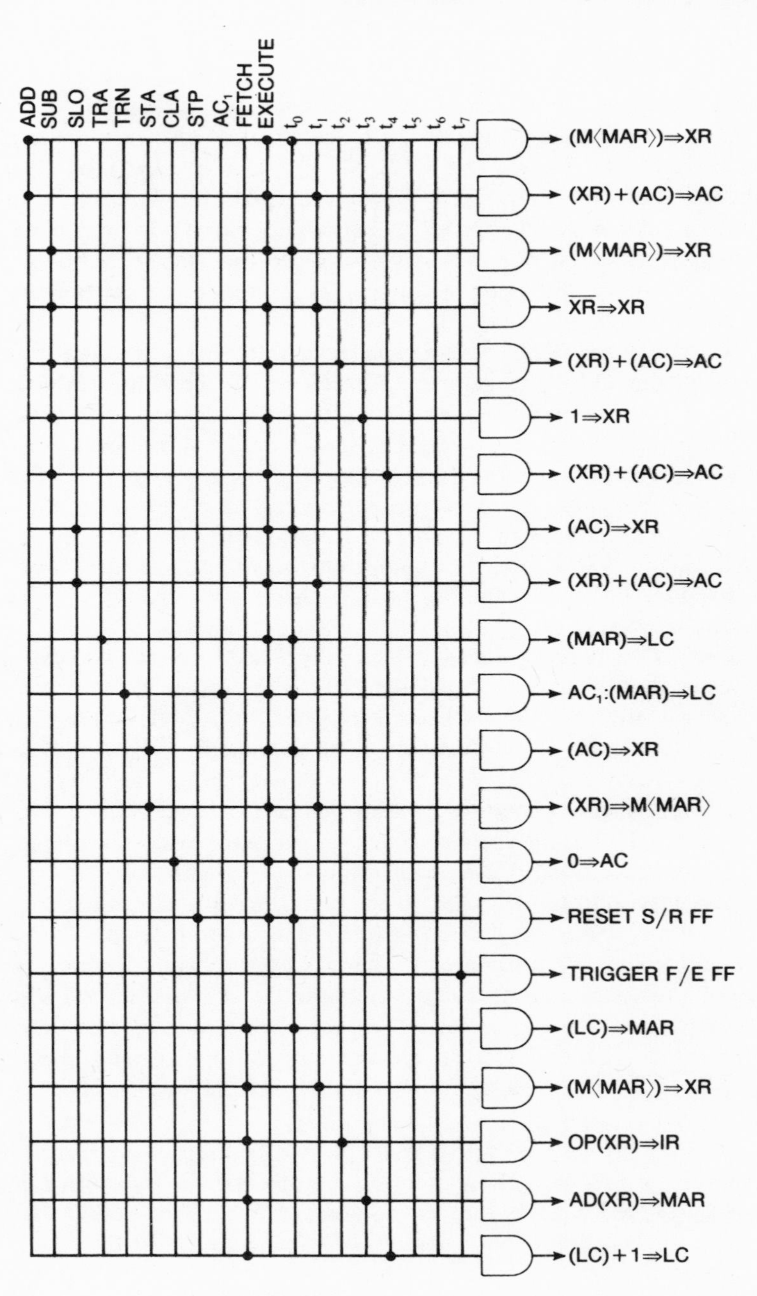

FIGURE 2-19 Subcommand Generator

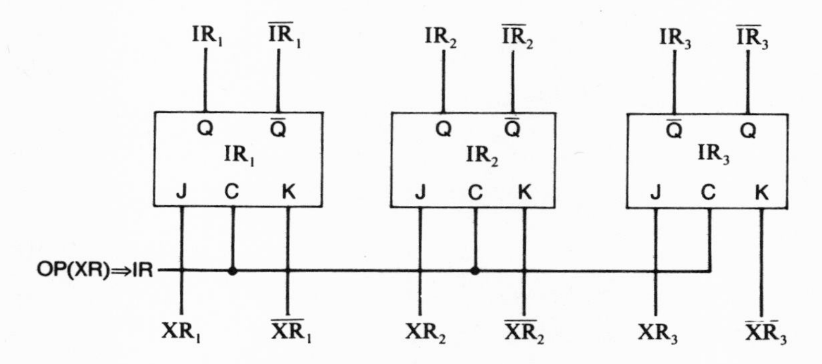

FIGURE 2-20 Instruction Register

Next, using the left side entries of the above list we identify each register in the WHYCO that is to be related to the exchange register. They are as follows:

1. XR
2. M<MAR>
3. AC

These lists tell us that a total of seven lines plus the clock pulse will need to be connected using the appropriate circuitry to the J and K terminals of the flip-flops which make up the exchange register.

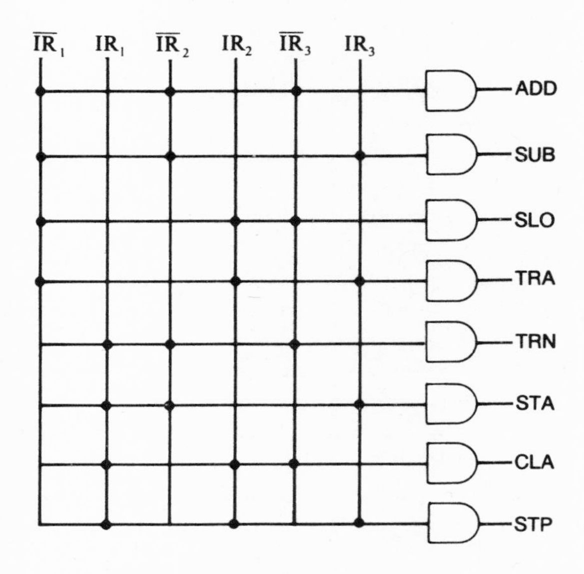

FIGURE 2-21 Instruction Register Decoder

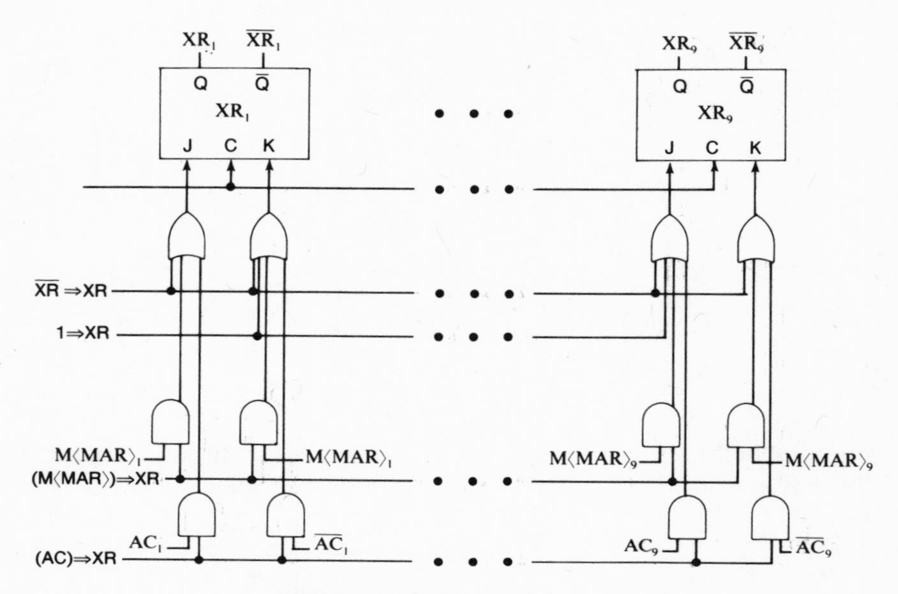

FIGURE 2-22 Exchange Register

The first subcommand $\overline{XR}\Rightarrow XR$ can actually be accomplished by complementing the existing values in the exchange register. Because this can be achieved by a simultaneous pulsing of the J and K terminals, there is no need to "reroute" the output lines of the exchange register back through as input. Thus, a total of six lines and the clock pulse will be sufficient.

Figure 2-22 is for the most part self-explanatory. However, let us analyze the implementation of one subcommand, namely, $(AC)\Rightarrow XR$. The reader may actually have more questions about $(M<MAR>)\Rightarrow XR$, but it would be advisable to return to these after the memory, memory address register, and memory address register decoder have been designed.

Obviously the accumulator cannot be connected to the exchange register by simply aligning the Q and $\overline{Q}$ lines of the accumulator with J and K terminals of the exchange register. Doing so would cause the exchange register to *always* match the accumulator, whereas we want the match to occur only when there is a one on the line $(AC)\Rightarrow XR$. However, ANDing the Q and $\overline{Q}$ outputs of the accumulator with the line $(AC)\Rightarrow XR$ will produce the desired result.

The reader is encouraged to study carefully Figure 2-22 as preparation for the design of the other registers in the WHYCO.

We will follow the same procedure for the **memory address register** (see Figure 2-23) that was used for the exchange register. First, we locate on the

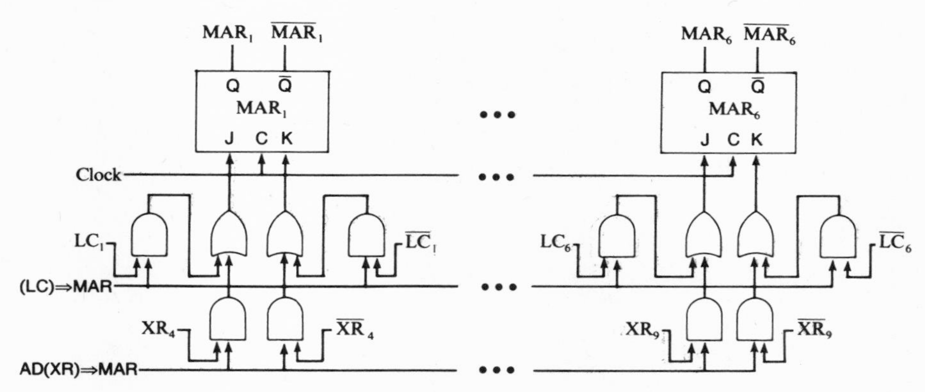

FIGURE 2-23 Memory Address Register

subcommand generator (Figure 2-19) every output line of the form ______⇒MAR. They are as follows:

1. (LC)⇒MAR
2. AD[XR]⇒MAR

Using the left side entries of this list, we identify each register that is to be related to the memory address register. They are as follows:

1. LC
2. XR

These lists tell us that a total of four lines plus the clock pulse will need to be connected using the appropriate circuitry to the J and K terminals of the flip-flops that make up the memory address register.

The "appropriate circuitry" can now be developed in a manner similar to that of the exchange register. It should be pointed out, however, that the

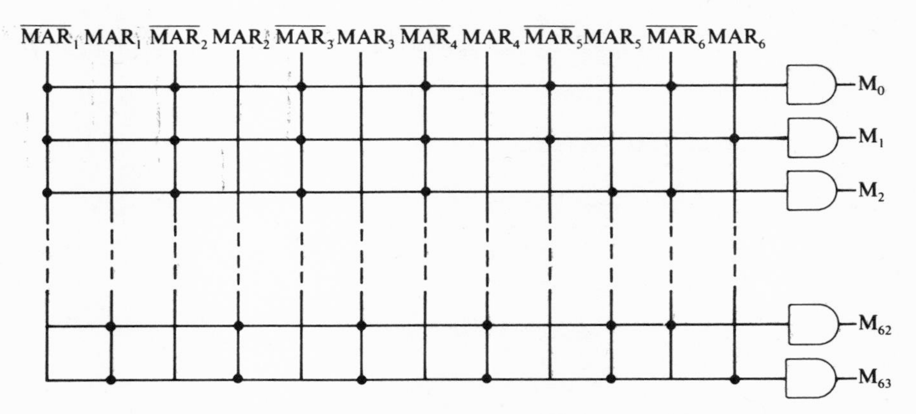

FIGURE 2-24 Memory Address Register Decoder

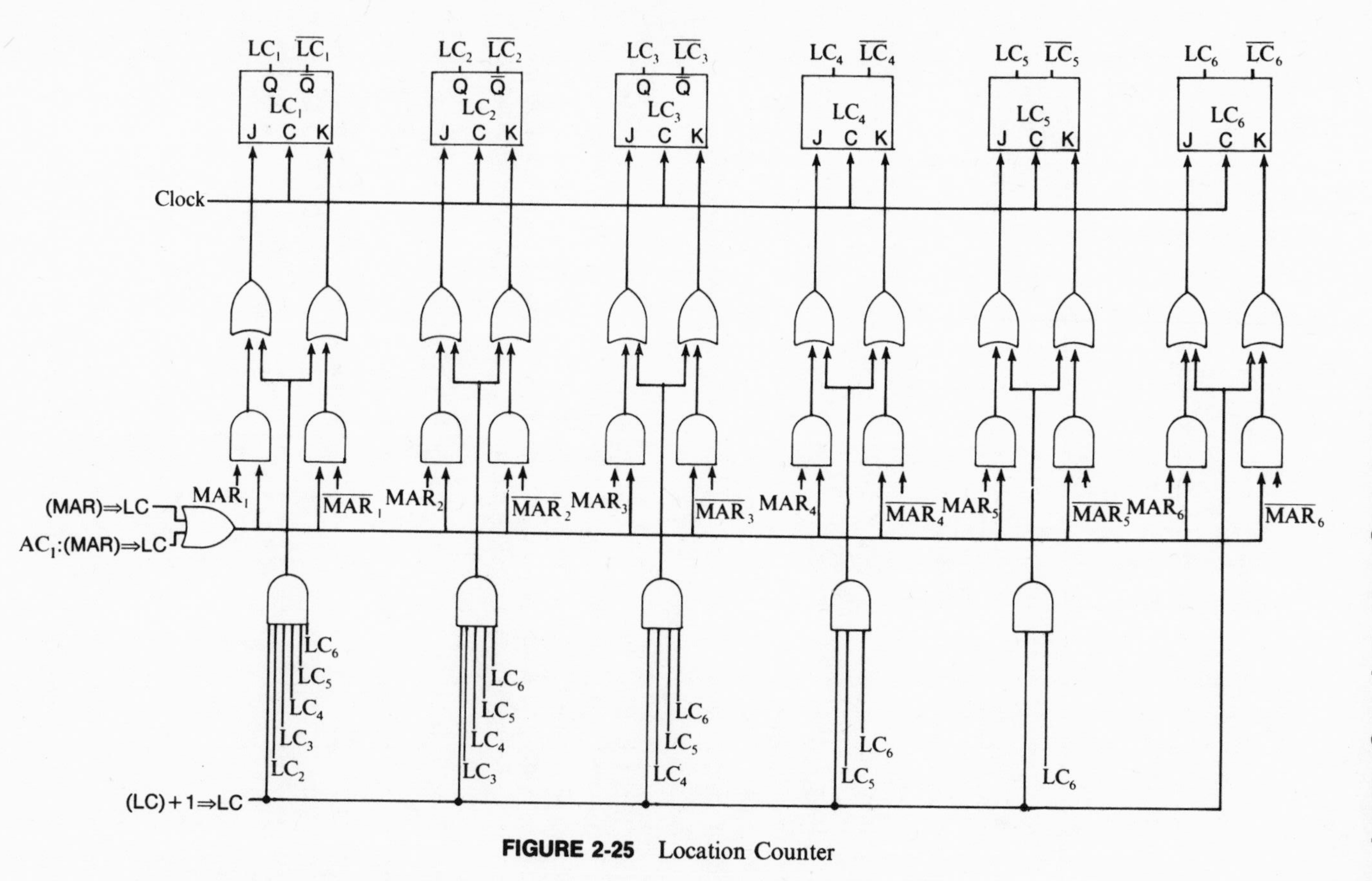

FIGURE 2-25 Location Counter

memory address register is only six bits long and the address portion of the exchange register is the last six bits. Therefore, XR_4 will be associated with MAR_1, XR_5 with MAR_2, and so on. The location counter, on the other hand, is only six bits long and therefore the bit positions will correspond with the bit positions in the memory address register.

For the **memory address register decoder** the method used here is exactly the same as the approach used when we designed the instruction register decoder. The output of the decoder (the AND gates on the right side of the single line diagram) will be a "one" on one of the lines, M_i, $i=0,1,\ldots,63$. These lines are used in the memory unit to select the memory location which is currently being referenced. Figure 2-24 is an abbreviated diagram for the memory address register decoder.

The **location counter** presents some design problems that are not incurred in any other register in the WHYCO. It must count as well as record. Let us deal with the latter function first. Once again we follow the same procedure. From the subcommand generator (Figure 2-19) we find the following output to the location counter:

1. **(MAR)⇒LC**
2. AC_1**:(MAR)⇒LC**

Using the left side entries we find that only the memory address register is involved in these subcommands. Consequently, they can be ORed together

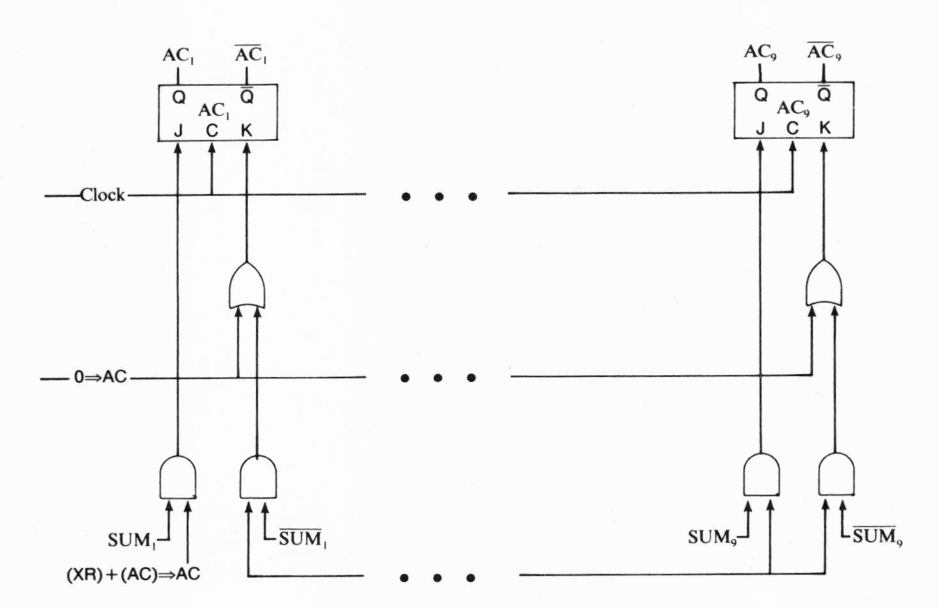

FIGURE 2-26 Accumulator

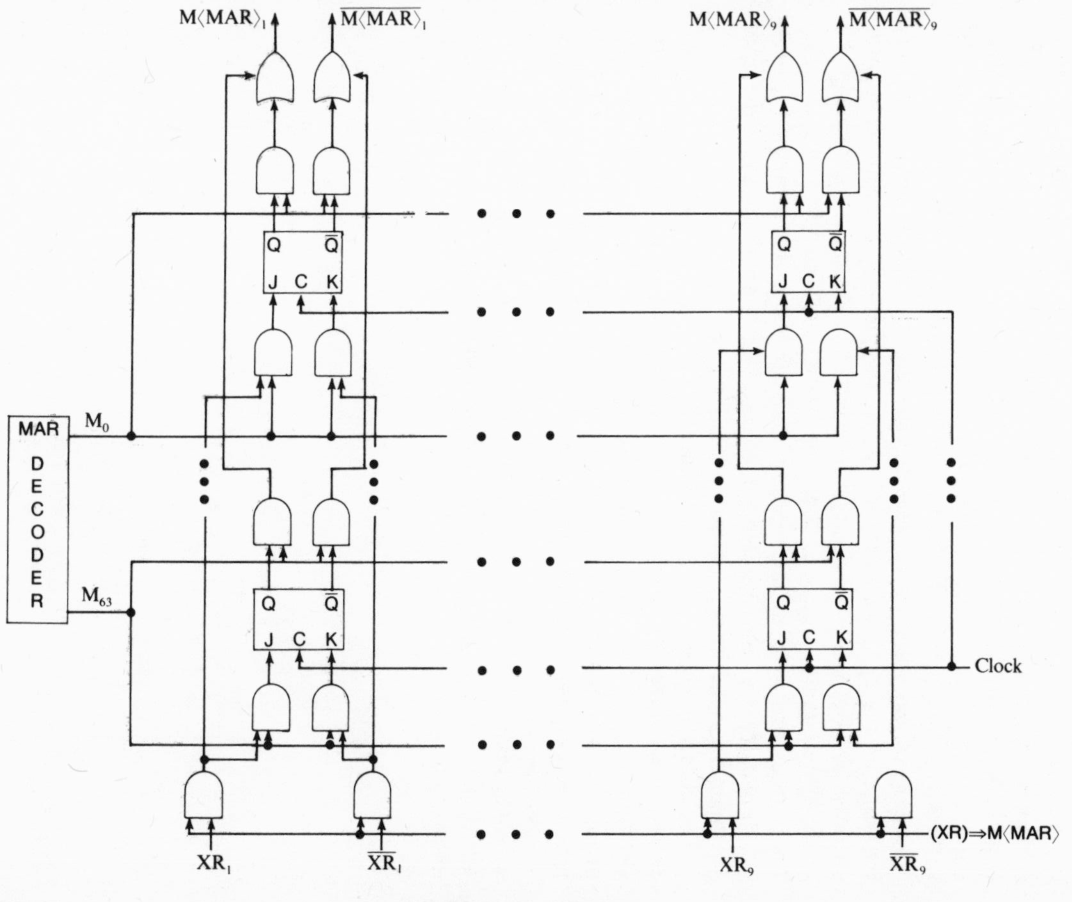

FIGURE 2-27 Memory

and then ANDed with the corresponding bit position in the location counter.

The counting function will be accomplished by complementing a bit position only when all the bit positions to its right are ones. Bit position six is an exception because it is the right-most flip-flop in the location counter. It will therefore be complemented each time (LC) + 1⇒LC has a one on it. ANDing all the outputs to the right of any flip-flop, and connecting the result to J and K terminals of that flip-flop, will accomplish the counting function of the location counter (see Figure 2-25).

It is hoped that the reader is now familiar and comfortable with the design procedures for the registers of the WHYCO, so we will not deal explicitly with these procedures as they relate to the **accumulator** but will, instead, provide only the final configuration (see Figure 2-26).

The **memory** design is best approached by assuming that it only contains a single register, say M_0. Here again the same design considerations apply and they are left as an exercise for the reader.

The primary conceptual difference lies in the fact that the memory is not a single register but rather 64 registers. The memory address register decoder determines which register will be referenced. This means that each J and K terminal will receive the ANDed results of XR_i and (XR)⇒M <MAR> and M_j. Also the Q and $\overline{Q}$ of each flip-flop must be ANDed with M_j and the result ORed with the corresponding result from the other registers in the memory to produce M<MAR>$_i$. Figure 2-27 shows the memory for M_0 and M_{63}.

2-8 Whyco Simulation

The simple computer that we have just designed is best appreciated by simulating its behavior either by hand or by another computer. In this section we will illustrate the simulation by hand and refer the interested reader to problem 1 of the Exercises.

The simulation is merely a matter of showing the changes that take place in each of the registers of the WHYCO as a program is being executed. To do this we need a sample program and its binary code. Consider the algorithm in Figure 2-28 which computes the absolute value of a difference.

In order to use this simple algorithm we must first make several assumptions, none of which detract significantly from our basic effort which is to simulate what occurs as the WHYCO executes this program.

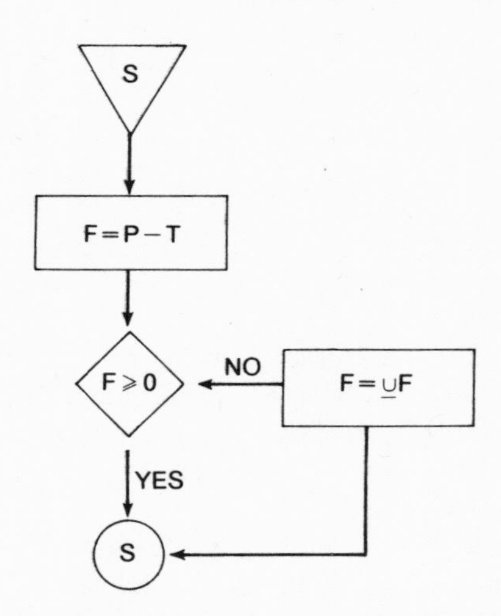

FIGURE 2-28 Flow Diagram for Absolute Value of a Difference

These assumptions are as follows:

1. The program is coded in binary and is in the WHYCO memory with the first program word in location 0.
2. The data is in the WHYCO memory with the first data element in location 30. Location 30 is arbitrary.
3. The final result will be in the accumulator.

All we are doing at this point is bypassing a very necessary feature of any computer, namely, input and output operations, commonly denoted by I/0.

We also encounter a second difficulty before we can trace through a simulation of our sample program. How do we get the program in binary code? Actually this question is answered in considerable detail in Chapters 3-8 of this text. Consequently, what we will do here will be a prelude to these chapters.

Table 2-1
Octal to Binary Conversion Table

OCTAL	BINARY
0	000
1	001
2	010
3	011
4	100
5	101
6	110
7	111

Table 2-2
Machine Code for Figure 2-28

MEMORY LOCATION:	CONTENT	COMMENTS
00:	CLA 00	0 in accumulator
01:	ADD 30	P in accumulator
02:	SUB 31	P-T in accumulator
03:	TRN 05	Transfer if accumulator is negative
04:	STP 00	Stop if accumulator is positive
05:	STA 32	Stores accumulator (F)
06:	CLA 00	0 in accumulator
07:	SUB 32	0-F in accumulator
10:	STP 00	Stop
30:	134	Location of $P=134_8$
31:	257	Location of $T=257_8$

It is convenient to consider the use of octal numbers at this point. Even though the WHYCO is a very small computer, it still becomes somewhat tedious to write out 9 bit words in binary form when the same task can be accomplished in 3-digit octal form. Octal coding is easy to use because it is a well-known fact that any three consecutive bits translate directly to a single octal digit and vice versa. The reader may wish to convince him or herself that this is the case. Table 2-1 will be useful.

A further coding convenience can be acheived by using the operation codes of the WHYCO in their mneumonic form. Thus our program with sample data might appear as in Table 2-2

The coding in Table 2-2 is very similar to the coding one would expect a compiler to produce. And, in fact, in Chapter 8 we will illustrate exactly how this code production takes place. For now, let us assume that the above coding is in the WHYCO in binary form. By referring to Figure 2-21 we can determine the binary codes for the operations of the WHYCO. They are listed in Table 2-3 as follows:

Table 2-3
Binary Operation Codes for Whyco

OPERATION:	BINARY CODE
ADD	000
SUB	001
SLO	010
TRA	011
TRN	100
STA	101
CLA	110
STP	111

Table 2-4 displays our program as it appears in the WHYCO memory.

Table 2-4
Binary Code for Figure 2-28

MEMORY LOCATION:	CONTENT
000000:	110000000
000001:	000011000
000010:	001011001
000011:	100000101
000100:	111000000
000101:	101011010
000110:	110000000
000111:	001011010
001000:	111000000
011000:	001011100
011001:	010101111

The trace of the execution of our program can be denoted by listing the various registers of the WHYCO as follows:

LC MAR XR IR AC

If the WHYCO begins in execute mode, some extraneous data will be processed and consequently we simply "wait" for t_7, at which time we enter the fetch mode with 00 in the location counter. By referring to Figure 2-19, the subcommand generator, we can write the trace of the FETCH cycle for the first instruction(Table 2-5).

Table 2-5
Fetch Cycle for First Instruction in Table 2-4

t_i	LC	MAR	XR	IR	AC
0	00	00	—	—	—
1	00	00	600	—	—
2	00	00	600	6	—
3	00	00	600	6	—
4	01	00	600	6	—

These values will remain unaltered during the remainder of the timing pulses, namely, t_5, t_6, and t_7. When t_7 occurs we will change to execute mode and continue processing. The program we are executing begins by clearing the accumulator to zero. This occurs at t_0 and leaves the following:

t_i	LC	MAR	XR	IR	AC
0	01	00	600	6	000

Once again we "wait" for t_7 to switch back to the FETCH cycle which will prepare us to execute the second instruction. Table 2-6 shows a partial trace for the sample program. It is a partial trace because we show only the results at the end (t_7) of the FETCH and EXECUTE cycles. As a result the t_i column has been retitled by F_n/E_n for FETCH word n or EXECUTE word n.

Table 2-6
Trace showing results of fetch and execute cycles for program in Table 2-4

F_n/E_n	LC	MAR	XR	IR	ACC
F_0	01	00	600	6	—
E_0	01	00	600	6	000
F_1	02	30	030	0	000
E_1	02	30	134	0	134
F_2	03	31	131	1	134
E_2	03	31	001	1	655
F_3	04	05	405	4	655
E_3	05	05	405	4	655
F_5	06	32	532	5	655
E_5	06	32	655	5	655
F_6	07	00	400	4	655
E_6	07	00	400	4	000
F_7	10	32	132	1	000
E_7	10	32	001	1	123
F_{10}	11	00	700	7	123
E_{10}	11	00	700	7	123

The execution of the subtraction operator is perhaps the most complex. Table 2-7 details the full sequence for line E_2 in Table 2-6.

Table 2-7

F_n/E_n	LC	MAR	XR	IR	ACC
F_2	03	31	131	1	134
E_{2_0}	03	31	257	1	134
E_{2_1}	03	31	520	1	134
E_{2_2}	03	31	520	1	654
E_{2_3}	03	31	001	1	654
E_{2_4}	03	31	001	1	655

If we were to fill in the details for each operator, the trace in Table 2-6 would be much longer.

We are now ready to begin our development of a high-level programming language for the WHYCO.

EXERCISES

1. Program the following in WHYCO binary code and trace the execution of your program showing the results of the fetch and execute cycles at t_7. Your trace will be similar in format to Table 2-6.

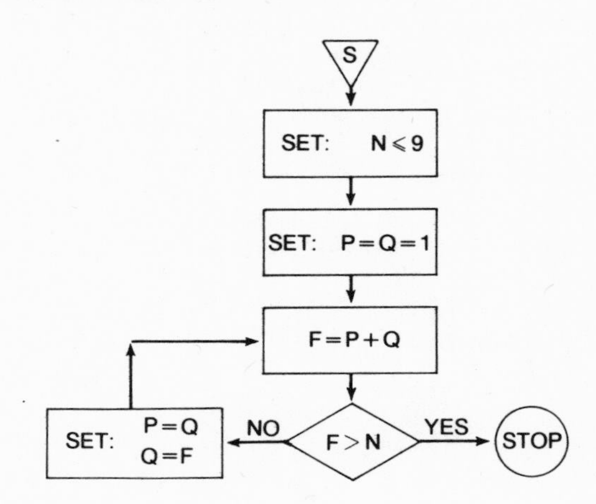

2. Show all the linkage between
   (a) The instruction register, the instruction register decoder, and the subcommand generator.
   (b) The subcommand generator and the exchange register.
   (c) The subcommand generator and the exchange register.
   (d) The subcommand generator and the memory address register.
   (e) The accumulator and the exchange register.
   (f) The exchange register and the memory address register.
3. Arithmetic units can be designed in a large variety of ways. One way is to use all half adders similar to the one designed in Figures 2-10 and 2-11. Design a 3-bit adder using only half adders.
4. The reason the timing counter is decoded into eight clock pulses is that the SUB instruction and the FETCH cycle contain five subcommands and the timing counter must produce 2^N pulses. In order to accomodate them we set $N=3$. The subcommand $\overline{XR} \Rightarrow XR$ is only used in the SUB instruction. Omit it and use $\overline{(M<MAR>)} \Rightarrow XR$ in subcommands for SUB instead of $(M\langle MAR\rangle) \Rightarrow XR$. In the FETCH cycle, transfer the instruction to the instruction register and the address to the memory address register at the same time. Trigger the FETCH-EXECUTE flip-flop on t_3. This will allow us to set $N=2$ and have a timing cycle of 4. Modify Figures 2-16, 2-18, 2-19, and 2-22 to show the implementation of these changes.

Chapter 3

High-Level Language Design

We traced through the computation process for a simple program in Section 8 of Chapter 2. The program, coded in binary and displayed in Table 2-4, can have several different representations. The representation in Table 2-4 is called the object code and as we have seen, can be converted to actual computation by the hardware. We also represented our program in Table 2-2 using the mnemonic form for the operations and octal numbers for the data and memory addresses. This representation is called the assembly code and is converted to object code by a program known as the assembler. In Figure 2-28 we supplied the flow chart for Table 2-2 and from it we could code the program in some high-level language such as FORTRAN, BASIC, or PL/1. This high-level representation is called the source code and can be converted to assembly code by a program known as a compiler.

There are other software devices such as translators and interpreters. Although these will not concern us in any direct way we mention them so as to define more accurately the role of the compiler and assembler. The term **translator** is often used to refer to a program that converts source code in one high-level language to source code in another high-level language. It has also been used as a general classification, that is, the term **translator** refers to a set which contains such things as compilers, assemblers, and interpreters. The term **interpreter** is more precise in that it always refers to a program which converts source code to computation and does not produce object code in the process. Figure 3-1 is an effort to display the terms we have been discussing in a diagrammatic fashion.

Our approach will begin at the top of Figure 3-1 and will develop all the mechanisms for compiling source code into assembly code. We will not continue the process by developing an assembler to convert the assembly code into object code. The reason for this omission is because an assembler for the WHYCO would be trivial, that is, it would convert each operation mnemonic to its octal equivalent and then convert all octal digits to three-place binary digits. That *is* the WHYCO assembler!

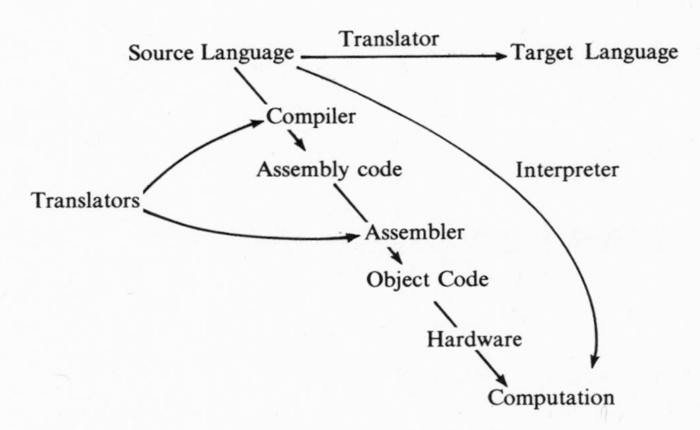

FIGURE 3-1 Source Language to Computation Scheme

The compiler is much more involved. Let us assume for the moment that some high-level language has been designed and we have used it to code a program. As far as the compiler is concerned such a program is nothing more than a long string of characters that must be read in one at a time. After they have been read, substrings must be recognized which constitute individual symbols. For example, a portion of a character string (program) may appear as follows:

---; IF, ---

In most high-level languages the substring IF is a symbol that signals the beginning of a conditional transfer of control in the program logic. The part of the compiler that recognizes such substrings is called the scan. After the scanning is completed we can classify all substrings as either operators or operands. This is the tabling function of the compiler. Most compilers READ, SCAN, and TABLE the input character string, but now we come to a fork in the road.

In the next section we will provide a formal definition of a high-level programming language. Such a definition consists of a set of rules which taken together constitute a syntax. Any program that is written according to these rules is said to be syntactically correct. In order to determine whether a program is syntactically correct the compiler must perform a syntax analysis. This analysis is called **parsing** and will be explained more fully in Chapter 6. The fork in the road has to do with the fact that it is possible to conduct the syntax analysis in at least the following ways:

1. Perform the syntax analysis as a separate function of the compiler.
2. Perform the syntax analysis as a parallel function with other functions of the compiler, such as code generation.

Although the second method is usually more efficient as far as actual applications are concerned, it is more difficult for the beginner to comprehend. We will, therefore, use the first method in our compiler.

Once we are assured that the program is syntactically correct, we reorganize the operators and operands into specialized data structures called stacks. Stacks, or push-down stacks as they are often called, will be defined in Chapter 7. From these stacks the compiler can generate the assembly code for the assembler and the compilation process is complete. Figure 3-2 shows a simple flow chart for the entire process.

At this point, we should no doubt point out that we are dealing initially with only the conceptual aspects for constructing a compiler. Were we to begin with specific details we would not be able to see the forest for the trees. For example, a character such as the letter "P" must be ultimately represented in binary code. Details such as this will be dealt with later on, and when they are, we will need to expand the memory of the WHYCO, and consequently, the word length. These are not difficult problems and the reader should set them aside for the moment until our general design considerations for the compiler are complete.

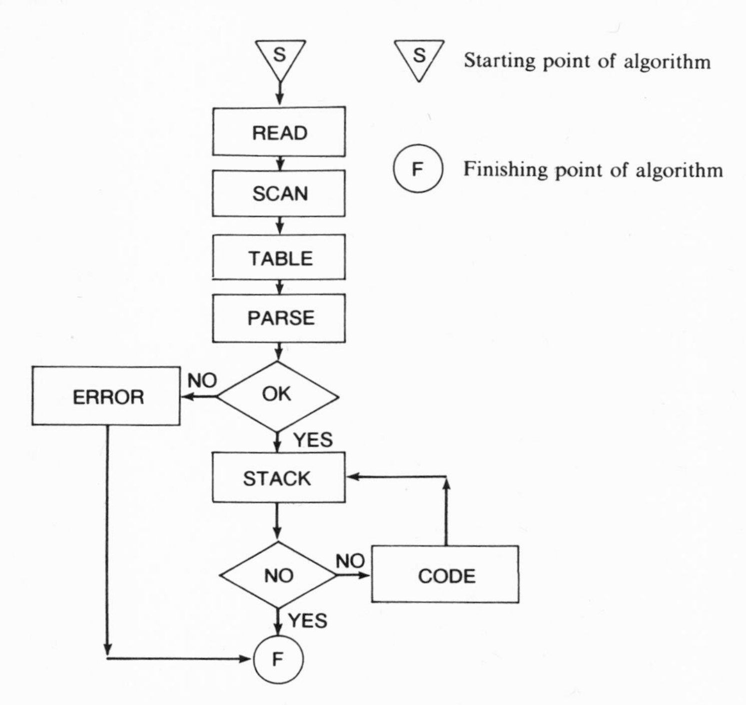

FIGURE 3-2 General Flow Chart for a Compiler

We will eventually add considerable detail to the flow chart presented in Figure 3-2, but first we must turn our attention to the high-level language which we intend to implement with our compiler. We have chosen to call this programming language PL/W.

3-1 A High-Level Language for the WHYCO (PL/W)

It would be very easy to spend a great deal of time designing a language for even our simple computer. It is also possible to save a great deal of time by illustrating the language and then, once we are reasonably familiar with the kinds of "sentences" that the language allows, formalize the structure of the language using some standard notation for the syntax. In order to provide such an illustration we will need a sample program.

3-2 A Sample Program

The Fibonacci sequence provides a good illustrative program which is simple but not trivial. Such a sequence requires two numbers with which to start. After that, each entry in the sequence is the sum of the previous two. Assuming that both of the initial numbers equal one, we obtain the following:

$$1,1,2,3,5,8,13,21,34,55,\ldots$$

Hopefully, the reader will recognize this sequence as an extension of the result obtained in exercise (1) of Chapter 2.

Consider the following code:

```
N=9;
P=1;
T=1;
F=P+T;
IF,F−N,↓1↓4;
P=T;
T=F;
GT ↑4;
SP;
ND
```

The first four statements require no explanation except possibly the semicolon. This character is being used simply to indicate the end of a statement. The fifth statement is a bit more involved. This statement is conditional transfer. In the event that $F-N$ is negative, it will transfer control one statement ahead in the program. In the event that $F-N$ is zero or positive, it will transfer control four statements ahead in the program. If either of the directionals had pointed upward, then control would have been transferred backward by that amount. The sixth and seventh statements are similar to the first four and consequently require no explanation. Statement eight is an unconditional transfer and in this program it transfers control four statements back. The ninth line is a STOP statement and the tenth is an END statement.

Readers with some programming experience will note that none of the statements in our sample program are labeled. This feature has been omitted because labeled statements introduce some tedious difficulties in the compilation process and it is best at the elementary level to avoid them.

The program on p. 44 is an example of PL/W and using it as a starting point we can now turn our attention to the syntactic definition of this language.

3-3 Syntax

The syntactic properties of a programming language are usually expressed in Backus-Naur form or BNF as it is commonly called. This form makes an important distinction between terminal and nonterminal symbols. All of the symbols on p. 44 are terminal symbols. The complete set of terminal symbols for PL/W is as follows:

TS = {1, 2, 3, 4, 5, 6, 7, 8, 9, D, F, P, T,
G, I, S, N, , , ; , $\underline{U}$, −, +, =, ↑, ↓, GT, IF, SP, ND}

The nonterminal symbols are defined in terms of the terminal symbols and in BNF notation they are always enclosed in wedge-shaped brackets. The complete set of nonterminal symbols for PL/W is as follows:

NS = {⟨PROGRAM⟩, ⟨STATEMENT⟩, ⟨ASSIGNMENT⟩,
⟨TRANSFER⟩, ⟨CONDITIONAL⟩, ⟨VARIABLE⟩,
⟨EXPRESSION⟩, ⟨INTEGER⟩, ⟨TERM⟩, ⟨ADDITIVE⟩,
⟨DIGIT⟩, ⟨DIRECTIONAL⟩}

The set *NS* and the set *TS* can now be combined to produce the PL/W syntax. The symbol : := can be read as "is defined by" and the symbol | can be read as "or." Table 3-1 is the set of production rules for PL/W. Statement (1) should be read as follows:

DEFINITION A program is defined by a statement followed by a semicolon followed by a program or a statement followed by a semicolon followed by ND.

In Table 3-1, the symbols IF, GT, SP, and ND represent the programming operations of IF, GO TO, STOP, and END, respectively. Observe that in the first part of the above definition the term **program** appears. This allows us to continue recursively to define program until the second alternative is used. Thus a program can be defined to contain an arbitrary number of statements.

Table 3-1
Syntax for PL/W

(1) ⟨PROGRAM⟩::= ⟨STATEMENT⟩; ⟨PROGRAM⟩|⟨STATEMENT⟩; ND

(2) ⟨STATEMENT⟩::= ⟨ASSIGNMENT⟩|⟨TRANSFER⟩|⟨CONDITIONAL⟩|SP

(3) ⟨ASSIGNMENT⟩::= ⟨VARIABLE⟩=⟨EXPRESSION⟩

(4) ⟨TRANSFER⟩::= GT ⟨DIRECTIONAL⟩⟨DIGIT⟩

(5) ⟨CONDITIONAL⟩::= IF, ⟨EXPRESSION⟩, ⟨DIRECTIONAL⟩ ⟨DIGIT⟩⟨DIRECTIONAL⟩⟨DIGIT⟩

(6) ⟨EXPRESSION⟩::= ⟨TERM⟩|⟨TERM⟩⟨ADDITIVE⟩⟨EXPRESSION⟩

(7) ⟨TERM⟩::= ⟨INTEGER⟩|⟨VARIABLE⟩| U⟨VARIABLE⟩

(8) ⟨INTEGER⟩::= ⟨DIGIT⟩|U⟨DIGIT⟩

(9) ⟨DIGIT⟩::= 1|2|3|4|5|6|7|8|9

(10) ⟨VARIABLE⟩::= D|F|G|I|N|P|S|T

(11) ⟨ADDITIVE⟩::= +|−

(12) ⟨DIRECTIONAL⟩::= ↑|↓

Later on, in Chapter 5, an even greater appreciation for the syntax provided in Table 3-1 will be realized. It is a simple syntax in that it does not provide any of the following:

1. I/O statements
2. labeled statements
3. multi-character variable names
4. floating point numbers
5. parenthetical expressions

and several other less obvious omissions. It is, however, more than adequate for our purposes.

EXERCISES

1. Just prior to Table 3-1 we gave an English translation of line (1) of the PL/W syntax. Write out a similar translation for lines (2), (3), (4), (5), and (6).

2. Given the following sets

   TS = {HEATHER, CHAD, SHANNAN, GABRIEL, OR, AND, IS, ARE, CHILDREN, PEOPLE, DELIGHTFUL, ENERGETIC, HAPPY, , , .}

   NS = {⟨SENTENCE⟩, ⟨SUBJECT⟩, ⟨PREDICATE⟩, ⟨PROPER NOUN⟩, ⟨NOUN⟩, ⟨VERB⟩, ⟨ADJECTIVE⟩, ⟨CONJUNCTION⟩}

   Use BNF notation to define all the elements of *NS*. Some sample sentences are:

   (a) Heather is delightful.
   (b) Heather, Chad, and Shannan are children.
   (c) Gabriel and Chad are energetic people.

Chapter 4

The Scan Function

Before we move on to the scan function, we must point out a number of things. First, the compiling process requires that the character string that comprises our program be read into the memory. Let us first look at our program as a string of characters. We have

N=9;P=1;T=1; F=P+T;IF,F−N,↓1↓4;P=T;T=F;GT↑4;SP;ND

When these characters are read into the computer they must be represented in some binary code. A number of such codes have been invented. Some of these are as follows:

1. BCD Binary Coded Decimal
2. EBCDIC Extended Binary Coded Decimal Interchange Code
3. ASCII American Standards Code for Information Interchange

There has been some standardization along this line, but none of the codes are completely standard and in some cases they are not particularly well suited to the compiler.

4-1 Internal Compiler Code

Due to these difficulties, the code is converted to an internal compiler code (ICC). Because we are not dealing with any input devices we can concentrate on the ICC and assume that when our characters are read they are stored in the memory in their ICC form. In Table 4-1 we provide the ICC for PL/W.

The program on p. 44 consists of 48 characters and would therefore occupy 48 of our 64 memory locations. Obviously, the memory of the WHYCO is too small for any compiler.

It should not be difficult to enumerate the necessary changes that would produce more memory. For example, adding a single bit position to the word length would allow us to increase the memory address to seven bits and thereby gain access to 64 additional memory locations. We would also

Table 4-1
Internal Compiler Codes

CHARACTER	OCTAL CODE	BINARY CODE
1	01	000001
2	02	000010
3	03	000011
4	04	000100
5	05	000101
6	06	000110
7	07	000111
8	10	001000
9	11	001001
D	12	001010
F	13	001011
P	14	001100
T	15	001101
G	16	001110
I	17	001111
S	20	010000
N	21	010001
,	22	010010
;	23	010011
␣	24	010100
—	25	010101
+	26	010110
=	27	010111
M	30	011000
↑	31	011001
↓	32	011010

have to alter all the registers to handle a ten-bit word. The memory address register decoder, location counter, and location counter decoder would also be changed so as to gain access to the additional memory. If a ten-bit word proves insufficient, we could add another bit position. The reader should have no difficulty in returning to Chapter 2 and identifying the appropriate changes.

Most computers have word lengths far greater than 9, 10, or 11. In fact, 32 is a very common length and characters that usually require eight bit positions for their representation can be stored four to a word.

In the WHYCO, however, we will not attempt to economize on our storage. Consequently the 48 locations containing our program will have at least three "wasted" bit positions because our word length is at least nine bits and our ICC is six bits. Using octal representation and assuming that the program was read in character by character, beginning with location 1, we have the locations and IC codes shown in Table 4-2.

Table 4-2
Character String Representation

CHARACTER	LOCATION	ICC
N	1	21
=	2	27
9	3	11
;	4	23
P	5	14
=	6	27
1	7	01
;	10	23
T	11	15
=	12	27
1	13	01
;	14	23
F	15	13
=	16	27
P	17	14
+	20	26
T	21	15
;	22	23
I	23	17
F	24	13
,	25	22
F	26	13
—	27	25
N	30	21
,	31	22
↓	32	32
1	33	01
↓	34	32
4	35	04
;	36	23
P	37	14
=	40	27
T	41	15
;	42	23
T	43	15
=	44	27
F	45	13
;	46	23
G	47	16
T	50	15
↑	51	31
4	52	04
;	53	23
S	54	20
P	55	14
;	53	23
N	57	21
D	60	12

This information, stored as we have indicated in Table 4-2, constitutes a character string. We are now ready to consider the task of the scanning function.

The purpose of the scan is to identify substrings of characters which should be treated as individual symbols. If we made the following replacements in the PL/W syntax, we could avoid the scanning function altogether:

1. Replace IF by I
2. Replace GT by G
3. Replace SP by S
4. Replace ND by N

In other words, the two-character operator symbols have been used in the PL/W syntax in order to force the need for the scanning function. This may appear arbitrary at first, but the reader should be reminded that PL/W is a pedagogical tool and to have omitted the scanning function would have amounted to omitting a very essential feature of any compiler.

It should be easy for anyone with a modest amount of programming experience to see the necessity of the scanning function. The most obvious reason as to why the scanning function is required involves numbers. In order to keep things simple, PL/W allows only single-digit numbers, but clearly, this would be totally unworkable for a "real" computer. There are other reasons besides numbers, however. For example, in the PL/W syntax we have allowed only eight single-letter variable names, whereas in most programming languages the programmer is free to create variable names as he or she sees fit. These variable names may not be totally arbitrary—for example, they usually must begin with an alphabetic character—but for the most part they are virtually unrestricted. A physicist may prefer variable names such as MASS, GRAMS, GAMMA, and so on, whereas the economist may prefer such names as REVENUE, PROFIT, SHORT RUN, LONG RUN, and so on. Names such as these not only perform the general function of identifying a value, but they also signify its meaning. Even if signification of meaning was not a consideration, one could easily imagine a computer program with more than 26 variables and therefore a restriction of the variable names to single letters would prevent the writing of such programs. Thus, the scanning function is a necessary part of a compiler. In the compiler for PL/W we will be content to illustrate the scanning function only with regard to the symbols IF, GT, SP, and ND.

Essentially the problem is one of converting a character string into a symbol string. Mathematicians have had a field day with this problem, and the best scanners are a result of their efforts. Because the mathematical approach to this problem would carry us well beyond the scope of this text, we will have to be content with simply illustrating a way to accomplish the scanning function for the PL/W compiler.

4-2 The PL/W Scan Function

In any computer, arrays must be linearized, and therefore, when we encounter an array of characters in the PL/W compiler which represent a single symbol we must devise a method for dealing with it in a linear fashion.

Before we begin the task of developing an algorithm which will perform the scanning function, we must decide how we will record the results of the scan. We have already pointed out that we are converting a character string into a symbol string. In the character string the individual characters are easily distinguished from one another because they each occupy a separate location in memory. Because this will no longer be the case in the symbol string, we must devise some other means for distinguishing a symbol from those on either side of it.

The device which we will employ consists of inserting a marker (M) between each symbol and at the beginning of the string. This device will nearly double the amount of memory required, and so the word length of the WHYCO will have to take on at least one additional bit position and the rest of the hardware altered accordingly.

In Table 4-1, we have designated 30 as a marker. After the scan is complete, our program will be stored sequentially in memory, only now the marker 30 will occupy a memory location between each symbol. For our example, the storage begins at location 61_8 (see Table 4-3).

Table 4-3
Scanned Character String

SYMBOL	LOCATION	ICC
M	61	30
N	62	21
M	63	30
=	64	27
M	65	30
9	66	11
M	67	30
;	70	23
M	71	30
P	72	14
M	73	30
=	74	27
M	75	30
1	76	01
M	77	30
;	100	23
M	101	30
T	102	15
M	103	30

Table 4-3 (Continued)

SYMBOL	LOCATION	ICC
=	104	27
M	105	30
1	106	01
M	107	30
;	110	23
M	111	30
F	112	13
M	113	30
=	114	27
M	115	30
P	116	14
M	117	30
+	120	26
M	121	30
T	122	15
M	123	30
;	124	23
M	125	30
{ I	126	17
{ F	127	13
M	130	30
,	131	22
M	132	30
F	133	13
M	134	30
−	135	25
M	136	30
N	137	21
M	140	30
,	141	22
M	142	30
↓	143	32
M	144	30
1	145	01
M	146	30
↓	147	32
M	150	30
4	151	04
M	152	30
;	153	23
M	154	30
P	155	14
M	156	30
=	157	27
M	160	30
T	161	15
M	162	30
;	163	23

Table 4-3 (Continued)

SYMBOL	LOCATION	ICC
M	164	30
T	165	15
M	166	30
=	167	27
M	170	30
F	171	13
M	172	30
;	173	23
M	174	30
{G	175	16
T	176	15
M	177	30
↑	200	31
M	201	30
4	202	04
M	203	30
;	204	23
M	205	30
{S	206	20
P	207	14
M	210	30
;	211	23
M	212	30
{N	213	21
D	214	12

The brackets at the left in Table 4-3 are merely for the purpose of displaying more vividly the two-character symbols in our program.

We can now turn our attention to the more interesting task of designing an algorithm which will take the characters on p. 44 as input and produce Table 4-3 as output. This is easily accomplished by recognizing that the insertion of a marker around each character whose IC code is greater than 21 will produce the desired result. Figure 4-1 contains such an algorithm and uses the following key:

KEY BC: Points to location in character string

BS: Points to location in symbol string

UNDERSCORE: $\underline{A}$ refers to the location in memory designated by the value of A

OVERSCORE: $\overline{A}$ refers to the value in the location designated by the value of A

In order to illustrate the coding for Figure 4-1 without specifying the exact locations in memory, let us adopt the convention of denoting memory locations for the scan function of the compiler by S_1, S_2, S_3, and so on.

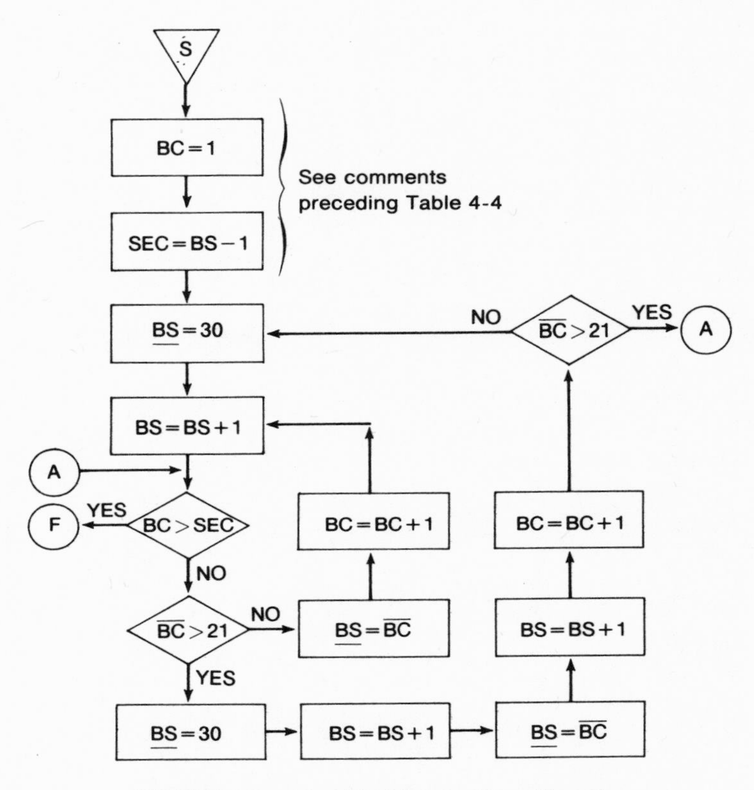

FIGURE 4-1 Algorithm for the Scan Function

The part of the coding that requires the most study has to do with the overscored and underscored variable names. For example, for the statement $\underline{\text{BS}}=30$ we could proceed as follows:

S_{12}: CLA
S_{13}: ADD "30"
S_{14}: STA $\underline{?}$

where "30" is merely a way to denote the memory address that contains the number 30. The real problem lies in the fact that $\underline{\text{BS}}=30$ means that 30 is to be stored in the location whose value is the value of the variable BS. How do we get that number into the rightmost 6+ bits of the operation in S_{14}? Suppose the following four operations were to precede the three listed above.

S_8: CLA
S_9: ADD S_{14}
S_{10}: ADD "BS"
S_{11}: STA S_{14}

Let us trace what will happen. At S_8 we will clear the accumulator to 000_8. In S_9 we add the contents of S_{14}, which must now be 500, to the accumulator. In S_{10} we add the contents of the memory location containing the value of BS. For our example program, the location for BS will be 61. The accumulator now contains 561. This result is now stored in S_{14}. In other words our program must "manufacture" the word in S_{14} just prior to executing it. We shall call any coding which prepares another code prior to its execution by the name "PREP code." Note that we are assuming a nine-bit word in our coding, whereas in actuality the WHYCO word will be increased to provide the memory required for the PL/W compiler.

The above coding will do just fine as long as the statement is not executed more than once. If it is, the second operation in the PREP codes will not be adding a word that contains an operation code followed by zeros, but rather, a word that contains the results of the previous execution of the PREP code.

The problem can be remedied in the following manner. We simply change the PREP code for S_{14} as follows:

S_8: CLA
S_9: ADD "I"
S_{10}: ADD "BS"
S_{11}: STA S_i

where I is the location that contains the appropriate initial value. In the above example I would contain 500. Our final coding for Figure 4-1 is given in Table 4-4.

Some additional comments regarding Figure 4-1 are in order. First, the initialization BC = 1 is no more than a reflection of the fact that we are assuming that the character string was read in beginning at location 1. In larger computers, this value would have been supplied by the operating system. Secondly, the statement SEC = BS − 1 stores the end of the character string. The value of BS will be the location immediately following the end of the character string. For our example program, BS = 61; the value of BS varies with different programs.

If we were to specify locations in WHYCO for BC, BS, SEC, and various constants such as 1, 500, 30, 100, and so on, we could complete the code using the PREP code method of Table 4-4. The reader should be reminded that the coding for the scanning function as well as all other parts of the PL/W compiler will occupy memory locations which cannot be used for character strings, symbol strings, tables, and other data structures. What we are alluding to here is a system problem in memory management and is not within the scope of our project. It is sufficient to simply point out that the data structures for the compiler must not be allowed to "erase" the compiler.

Table 4-4
Coding for the Scan Function

Memory Location:	Code	Figure 4-1	Reference
S_1:	CLA	BC = 1	
S_2:	ADD "1"		
S_3:	STA "BC"		
S_4:	CLA	SEC = BS − 1	
S_5:	ADD "BS"		
S_6:	SUB "1"		
S_7:	STA "SEC"		
S_8:	CLA	PREP CODE FOR S_{14}	
S_9:	ADD "500"		
S_{10}:	ADD "BS"		
S_{11}:	STA S_{14}		
S_{12}:	CLA	$\underline{BS} = 30$	
S_{13}:	ADD "30"		
S_{14}:	STA 00		
S_{15}:	CLA	BS = BS + 1	
S_{16}:	ADD "BS"		
S_{17}:	ADD "1"		
S_{18}:	STA "BS"		
S_{19}:	CLA	BC > SEC	(S_{22} can be completed when the first location of the table function is known)
S_{20}:	ADD "SEC"		
S_{21}:	SUB "BC"		
S_{22}:	TRN ☐		
S_{23}:	CLA	PREP CODE FOR S_{28}	
S_{24}:	ADD "000"		
S_{25}:	ADD "BC"		
S_{26}:	STA S_{37}		
S_{27}:	CLA	$\overline{BC} > 21$	
S_{28}:	ADD 00		
S_{29}:	SUB "22"		
S_{30}:	TRN S_{70}		
S_{31}:	CLA	PREP CODE FOR S_{37}	
S_{32}:	ADD "500"		
S_{33}:	ADD "BS"		
S_{34}:	STA S_{37}		
S_{35}:	CLA	$\underline{BS} = 30$	
S_{36}:	ADD "30"		
S_{37}:	STA 00		
S_{38}:	CLA	BS = BS + 1	
S_{39}:	ADD "BS"		
S_{40}:	ADD "1"		
S_{41}:	STA "BS"		

Table 4-4 (Continued)

MEMORY LOCATION:	CODE	FIGURE 4-1	REFERENCE
S_{42}:	CLA		
S_{43}:	ADD "000"	PREP CODE FOR S_{51}	
S_{44}:	ADD "BC"		
S_{45}:	STA S_{21}		
S_{46}:	CLA		
S_{47}:	ADD "500"	PREP CODE FOR S_{52}	
S_{48}:	ADD "BS"		
S_{49}:	STA S_{52}		
S_{50}:	CLA		
S_{51}:	ADD 00	$\underline{BS} = \overline{BC}$	
S_{52}:	STA 00		
S_{53}:	CLA		
S_{54}:	ADD "BS"	BS=BS+1	
S_{55}:	ADD "1"		
S_{56}:	STA "BS"		
S_{57}:	CLA		
S_{58}:	ADD "BC"	BC=BC+1	
S_{59}:	ADD "1"		
S_{60}:	STA "BC"		
S_{61}:	CLA		
S_{62}:	ADD "000"	PREP CODE FOR S_{66}	
S_{63}:	ADD "BC"		
S_{64}:	STA S_{66}		
S_{65}:	CLA		
S_{66}:	ADD 00		
S_{67}:	SUB "22"	$\overline{BC} > 21$	
S_{68}:	TRN S_8		
S_{69}:	TRA S_{19}		
S_{70}:	CLA		
S_{71}:	ADD "000"	PREP CODE FOR S_{79}	
S_{72}:	ADD "BC"		
S_{73}:	STA S_{79}		
S_{74}:	CLA		
S_{75}:	ADD "500"	PREP CODE FOR S_{80}	
S_{76}:	ADD "BS"		
S_{77}:	STA S_{80}		
S_{78}:	CLA		
S_{79}:	ADD 00	$\underline{BS} = \overline{BC}$	
S_{80}:	STA 00		
S_{81}:	CLA		
S_{82}:	ADD "BC"		
S_{83}:	ADD "1"	BC=BC+1	
S_{84}:	STA "BC"		
S_{85}:	TRA S_{15}		

EXERCISES

1. In the coding for Figure 4-1 we indicated the memory location for variables and constants by placing quotation marks around the variables and constants. Assume the following locations have been assigned:

VARIABLE/CONSTANT	LOCATION
1	500
BC	501
BS	502
SEC	503
500	504
30	505
000	506
21	507

Rewrite the contents of the following memory locations in octal and binary code.

$$S_2, S_3, S_5, S_6, S_7, S_9, S_{13}, S_{24}, S_{29}$$

Example solution for S_2

The octal code for ADD is 0 (see Table 2-3). Therefore we have:

$$S_2: 0500$$

Using Table 2-1, we can translate 0500 to binary. We obtain:

$$S_2: 000101000000$$

2. The technique employed in Figure 4-1 was to place a marker around each character whose IC code was greater than 21. Suppose we had attempted to implement the scanning function by placing a marker around each character whose IC code was less than 22. Why would such a technique not work?
3. The IF statement in the Fibonacci sequence program is as follows:

IF, F − N,↓1↓4;

If we simply placed a marker around each character whose ICC was greater than 21, we would obtain

IFM, MFM − MNM, MM↓M1M↓M4

Indicate which portion of the algorithm in Figure 4-1 avoids the double markers between the , and ↓.

4. Redefine the syntax in Table 3-1 so as to provide for multi-character integers and multi-character variables. Require that the multi-character variables begin with a letter and consist only of letters and digits. Does the algorithm in Figure 4-1 require modification in order to accomodate these syntactic changes? Using this revised syntax, design and code a PL/W program which will determine if a particular integer is a prime number.

Chapter 5

The Table Function

Here we are concerned not only with distinguishing an operator from an operand but also with tabling various characteristics of each. In PL/W we will need to construct a relatively simple table of operands. This simplicity is due to the fact that the PL/W syntax allows only eight variable names and all PL/W variables are integers. Consider the table function for operands in a much more complicated high-level language such as BASIC, FORTRAN, or PL/1. In these languages the number of different variable names is virtually unbounded and they can be of different types such as floating point variables, integer variables, character string variables, array variables, and so forth. The compilers for these kinds of languages must identify all of the variable names in any particular program and the program must also contain instructions for each variable as to its type. Sometimes these instructions are implicit as, for example, the convention in some versions of FORTRAN which allows all variable names beginning with I, J, K, L, M, or N to be treated as integer variables unless the program explicitly defines them to be floating point.

5-1 The PL/W Operand Location Table

Although the problem of tabling the characteristics of all the operands need not concern us, we nevertheless must be concerned with their location. The task is conceptually a simple one, namely, we step through the symbol string and record every operand and assign it a location. Implementation of this idea, however, is a bit more complicated. The difficulty lies in the fact that an operand may occur more than once in the symbol string and thereby be recorded more than once in the operand location table. We must therefore check the table for a given operand before we enter it in the table. These concepts are organized into an algorithm in Figure 5-1.

We shall now proceed to add more detail to Figure 5-1. First, we need to know the following numbers: (1) the beginning location of the symbol

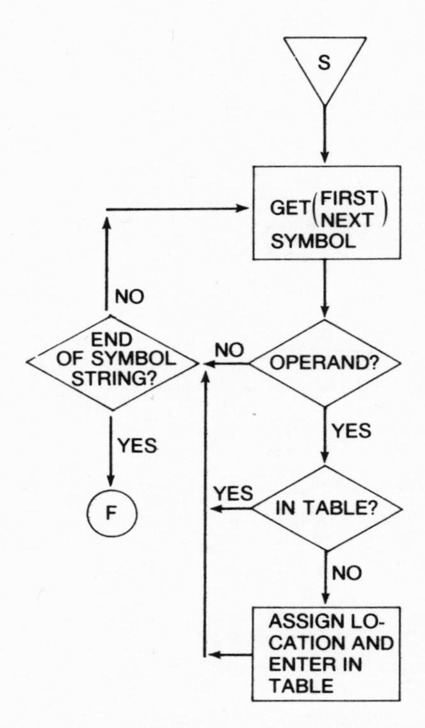

FIGURE 5-1 General Flow Chart for the Operand Location Table

string, that is, the initial value of BS, and (2) the beginning location for the operand location table. Let us discuss these in reverse order.

The beginning location of the operand location table could be almost anywhere. It seems logical, however, to have it follow the symbol string in memory. Note that in Figure 4-1 we exit at F after having incremented BS. In the case of our example program, BS would be 215 (see Table 4-3). Therefore, we can use the final value of BS from the scan function for the beginning location of the operand location table.

The beginning location of the symbol string has also been provided by the scan function as the final value of BC. Careful examination of Figure 4-1 reveals that we exit when BC>SEC, that is, when BC=61. The reader is reminded that values such as "61" are linked directly to our particular example program and, in general, would be provided by the input function which produces the character string and thereby records its final location.

We probably should mention that the scan function was designed several times before the author arrived at the final algorithm in Figure 4-1. The fact that the scan function provides values needed for the operand location table is not only not an accident, but also, required several revisions before it did so. Programmers, especially student programmers, should not be discouraged when they find that they cannot come up with the proper algorithm for a particular process on the first or even second effort.

Getting the first or next symbol requires that we check initially for a marker. Because the PL/W syntax allows only single-digit numbers and single-character variables, we can assume that the absence of the marker means one of two things. Either we have incurred one of our two character operators or we have an error in the character string. Because it is the task of the parse function to detect such errors we will simply continue to advance the value of BC until we incur a marker. This will mean that the I of IF, the G of GT, the S of SP, and the N of ND will end up on our operand table. This will not cause any problems, especially when one remembers that I, G, S, and N can be legitimate variable names and therefore operands in a PL/W program.

With the above considerations in mind, we can expand the GET (FIRST/NEXT) SYMBOL block and the END OF SYMBOL STRING diamond of Figure 5-1 into the following:

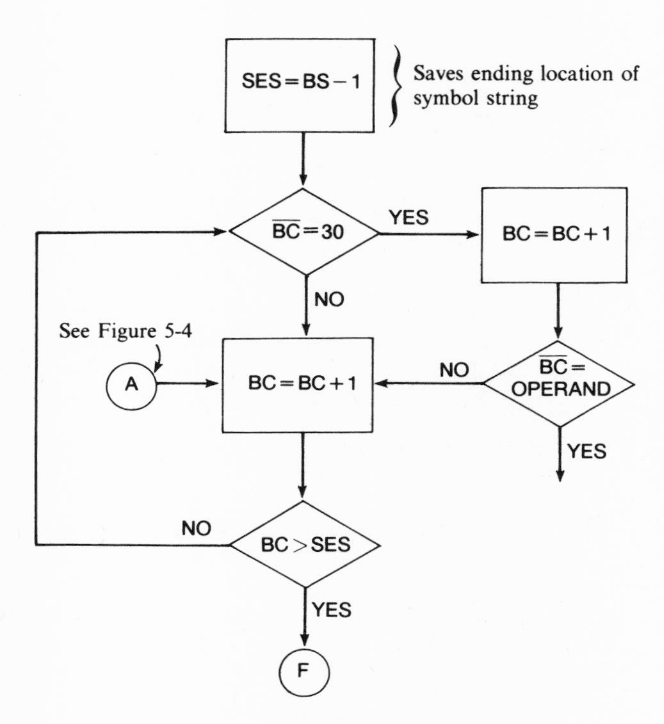

FIGURE 5-2

Our next task involves testing $\overline{BC}$ to determine whether or not we have incurred an operand. By examining the ICC in Table 4-1 we see that operands have ICCs that obey the following relationship:

$$\overline{BC} < 22$$

As a result, the $\overline{BC}$=OPERAND diamond in Figure 5-2 can be expanded as follows:

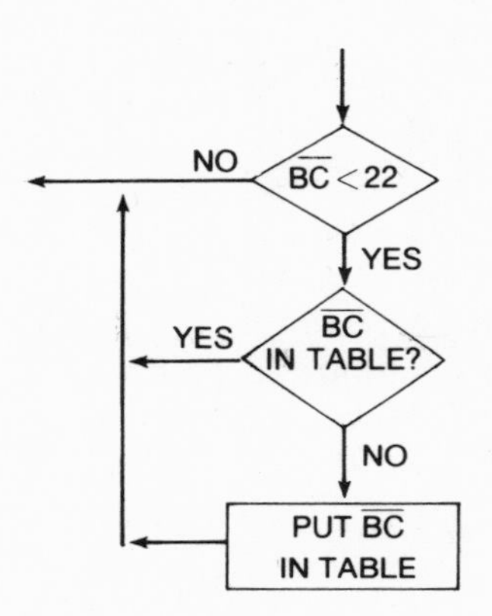

FIGURE 5-3

We now come to what will undoubtedly be a new concept for the beginner. What would the operand location table for PL/W look like if we were to write it down "on paper" as opposed to recording it in the computer's memory? It would have to contain the operand's name and location and, the location would have to contain the value of the operand, providing the value is known. The "new concept" concerns the fact that a number such as 9 will express two entirely different concepts. On the one hand, it will express a value, which is what one would expect, whereas on the other hand it will express a name for a value. This difference is not quite as uncommon as it may appear. Consider, for example, the fact that in everyday language we use the symbol "9" for the value and the symbol "nine" for the name. The "on paper" version of the operand location table for the Fibonacci sequence program is given in Table 5-1.

Because the operand location table must itself be located somewhere in the WHYCO memory, let us describe the table as it will appear "in memory." The table will begin in location 215; 215 will contain the ICC for N, 216 will be the location of N. In other words, each operand will be followed by its location in memory, and, in the event that the value of the operand is known, that location must contain that value. The "in memory"

Table 5-1
Operand Location Table

NAME	LOCATION	VALUE
N	216	—
9	220	9
P	222	—
1	224	1
T	226	—
F	230	—
4	232	4

Table 5-2
Representation of Operand Location Table

MEMORY LOCATION:	CONTENTS	EXPLANATION
215:	21	ICC for N
216:	—	
217:	11	ICC for 9
220:	11	Value of 9
221:	14	ICC for P
222:	—	
223:	1	ICC for 1
224:	1	Value of 1
225:	15	ICC for T
226:	—	
227:	13	ICC for F
230:	—	
231:	4	ICC for 4
232:	4	Value of 4
233:	30	Marker for end of table

results of the table functions for operands, or in other words, the operand location table, is given in Table 5-2.

With Table 5-2 to guide us we can now return to Fig. 5-3 and fill in the details. First, the check to see if $\overline{\text{BC}}$ is in the table will require that the table be initialized by placing the marker M in location 216. This will appear in our final flow chart as follows:

$$\underline{\text{BS}} = 30$$

For clarity we use the variable TL (table location) which must be initialized and saved. These operations will appear as follows:

TL = BS
STL = TL

We are now in a position to flow chart the check for the current $\overline{BC}$, and, the part which places $\overline{BC}$ in the table. This is done in Figure 5-4. Note that when the variable TL is advanced by 2 we "skip over" the location which contains the value of the operand.

The complete flow chart for the part of the table function which produces the operand location table is provided in Figure 5-5.

The actual coding for the operand location table will be very similar to the coding for the scan function that appears in Table 4-4. We will follow a

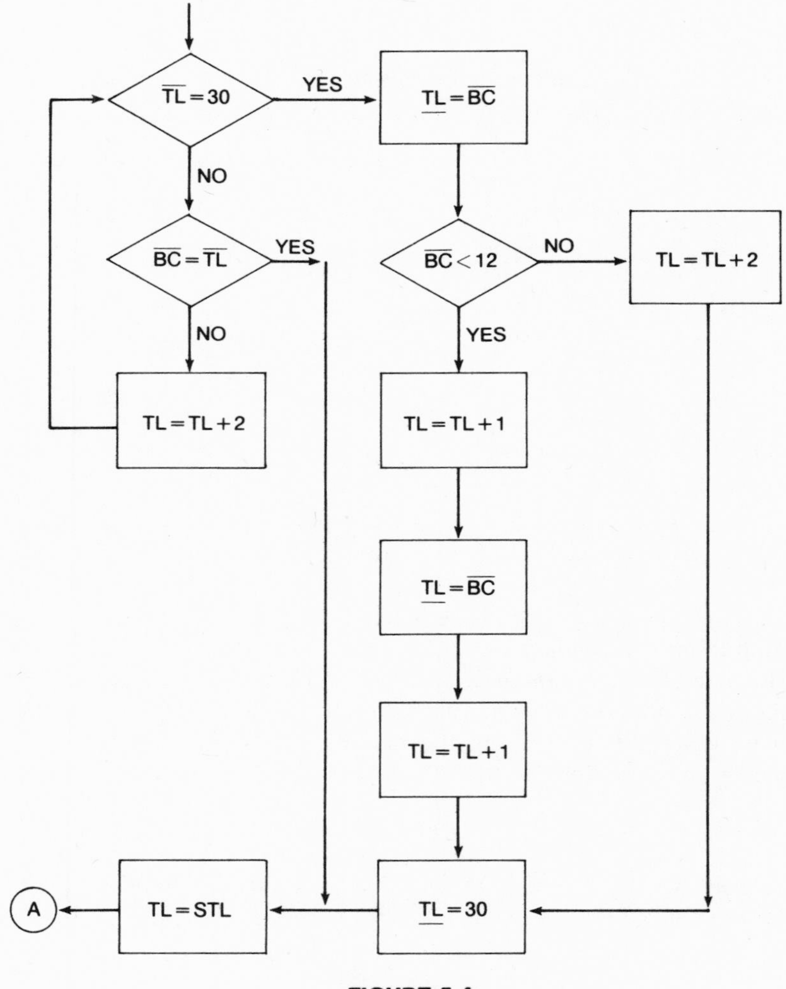

FIGURE 5-4

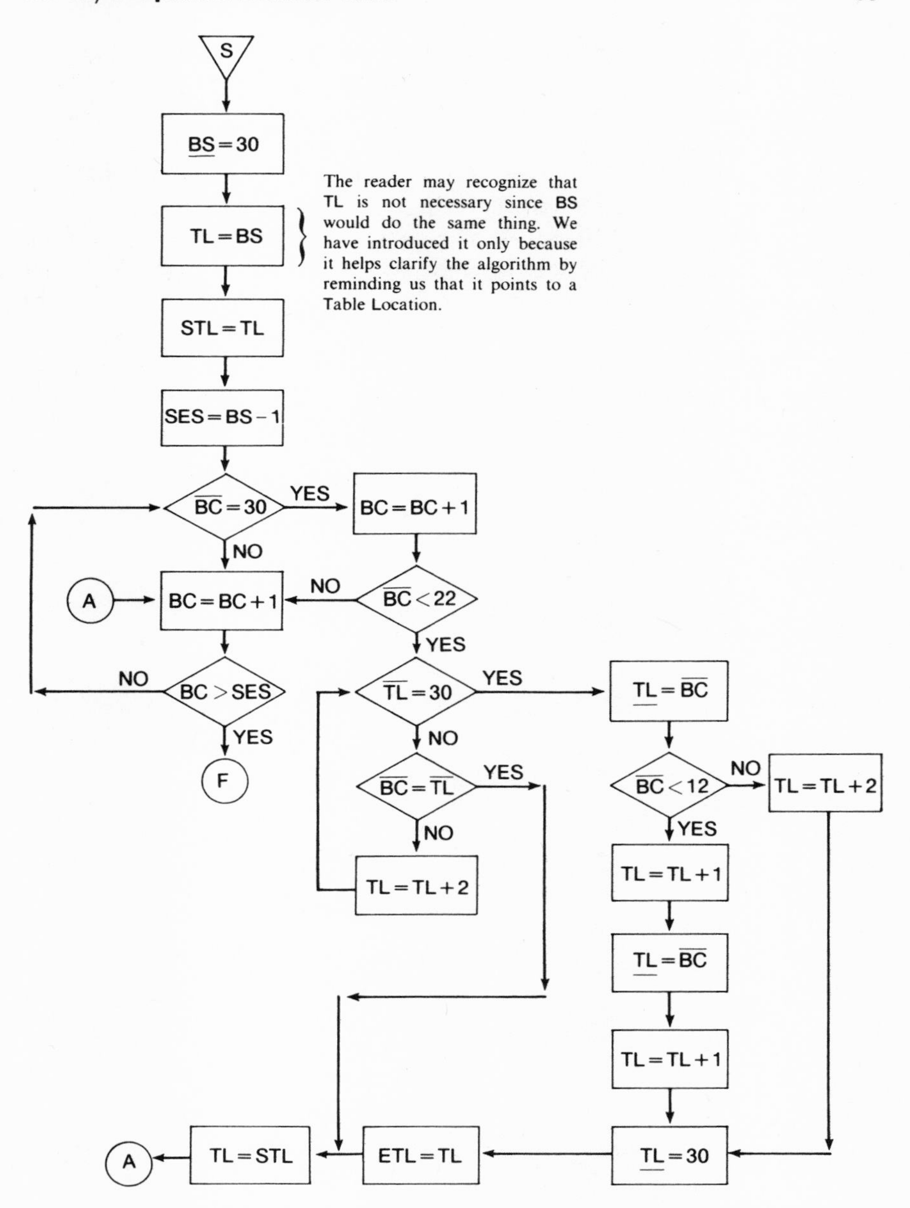

FIGURE 5-5 Algorithm for Operand Location Table Function

similar approach in regard to the operand location table but code only the portion of Figure 5-5 that appears in Figure 5-6.

We will assume that by now the reader is comfortable with the concept of PREP coding and proceed to supply in Table 5-3 the code for Figure 5-6, with no further explanation other than pointing out that the memory locations for the table function will be designated by a subscripted T.

Operators present a different set of circumstances altogether. First of all, they are not "created" by the program and therefore the compiler could contain a table of all the operators and their type. When we come to the STACK function for the PL/W compiler, we will need such a table for the PL/W operators. Before constructing this table let us turn our attention to the types of PL/W operators.

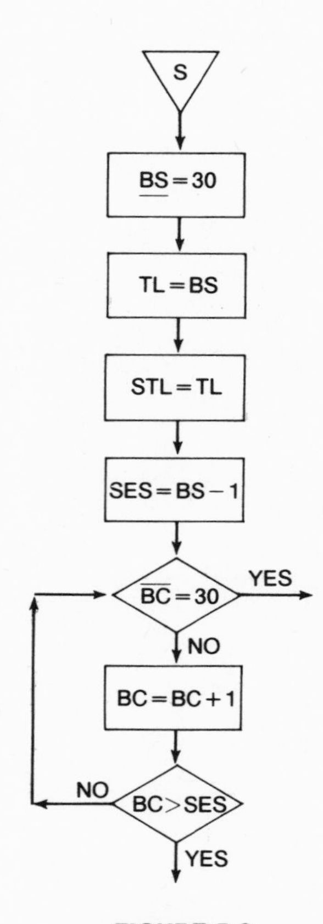

FIGURE 5-6

Table 5-3
Partial Coding for the Operand Location Table Function

MEMORY LOCATION:	CODE	FIGURE 5-6	REFERENCE
T_1:	CLA	PREP CODE FOR T_7	
T_2:	ADD "500"		
T_3:	ADD "BS"		
T_4:	STA T_7		
T_5:	CLA	$\underline{\text{BS}}=30$	
T_6:	ADD "30"		
T_7:	STA 00		
T_8:	CLA	TL = BS	
T_9:	ADD "BS"		
T_{10}:	STA "TL"		
T_{11}:	STA "STL"	STL = TL	
T_{12}:	SUB "1"	SES = BS − 1	
T_{13}:	STA "SES"		
T_{14}:	CLA	PREP CODE FOR T_{20}	
T_{15}:	ADD "100"		
T_{16}:	ADD "BC"		
T_{17}:	STA T_{20}		
T_{18}:	CLA	$\overline{\text{BC}} > 30$	
T_{19}:	ADD "30"		
T_{20}:	SUB 00		
T_{21}:	TRN ☐	TO [BC = BC + 1]	See note at T_{30}
T_{22}:	CLA	PREP CODE FOR T_{27}	
T_{23}:	ADD "000"		
T_{24}:	ADD "BC"		
T_{25}:	STA T_{27}		
T_{26}:	CLA	$\overline{\text{BC}} < 30$	
T_{27}:	ADD 00		
T_{28}:	SUB "30"		
T_{29}:	TRN T_{31}		
T_{30}:	TRA ☐	TO [BC = BC + 1]	This represents a different incrementation of the variable ***BC*** than that which is implemented at T_{31} (see Fig. 5-2).
T_{31}:	CLA	BC = BC + 1	
T_{32}:	ADD "BC"		
T_{33}:	ADD "1"		
T_{34}:	STA "BC"		
T_{35}:	CLA	BC > SES	
T_{36}:	ADD "BC"		
T_{37}:	SUB "SES"		
T_{38}:	TRN T_{18}		
T_{39}:	TRA ☐	TO PARSE FUNCTION	

5-2. The PL/W Operator Type Table

The reader is probably familiar with binary operators. A binary operator is one that requires two operands. Plus and minus are two such operators. Following this line of thought we have adopted the following type codes for PL/W operators (see Table 5-4).

Table 5-4
Operator Types

CODE	TYPE	MEANING
0	Null	Requires no operands
1	Unary	Requires one operand
2	Binary	Requires two operands
3	Tertiary	Requires three operands

Using these codes as well as the ICC, the second and fourth columns of Table 5-5 must be stored in the WHYCO memory.

Table 5-5
Operator Type Table

OPERATOR	OPERATOR ICC	TYPE	TYPE CODE
,	22	Null	0
;	23	Null	0
$\underline{\text{U}}$	24	Unary	1
—	25	Binary	2
+	26	Binary	2
=	27	Binary	2
GT	16, 15	Unary	1
IF	17, 13	Tertiary	3
SP	20, 14	Null	0
ND	21, 12	Null	0

Thus far we have used the first 60_8 locations in WHYCO for the input character string of the Fibonacci sequence program, and then 61_8 through 214_8 for the symbol string. We have also indicated that a large amount of memory would be devoted to coding scan and table functions. Assuming that further manipulation of the program input can be done without using memory locations beyond 277_8, let us load tables such as the one in Table 5-5 in the WHYCO memory beginning with location 300_8. Note that we will have to make use of the marker M (30) once again in order to identify two-character operators from one-character operators. We will also use the marker to separate the operator ICC from the type code for that operator.

In Chapter 7, where we design the PL/W stack function, the reader will discover that the operator-type table will never be searched for the "," or

";". These pseudo-operators are sometimes referred to as delimiters. Careful examination of the logic in Figure 7-11 will reveal the reasons for omitting these two symbols from the operator-type table. Also, the directionals can be considered as pseudo-operators because they will never be used for code generation. Therefore, they will not appear on the operator-type table. The PL/W operator type table is provided in Table 5-6.

Table 5-6
Representation of Operator Type Table

SYMBOL	LOCATION	ICC
M	300	30
U	301	24
M	302	30
Unary	303	1
M	304	30
–	305	25
M	306	30
Binary	307	2
M	310	30
+	311	26
M	312	30
Binary	313	2
M	314	30
=	315	27
M	316	30
Binary	317	2
M	320	30
G	321	16
T	322	15
M	323	30
Unary	324	1
M	325	30
I	326	17
F	327	13
M	330	30
Tertiary	331	3
M	332	30
S	333	20
P	334	14
M	335	30
Null	336	0
M	337	30
N	340	21
D	341	12
M	342	30
Null	343	0
M	344	30

EXERCISES

1. The following program is written in FORTRAN. It is designed to produce the Fibonacci sequence. List the operands and operators. [*Hint*: The best definition for an operator in this setting is "an operator is *not* an operand."]

```
   INTEGER    A,B,C
   A=1
   B=1
10 C=A+B
   WRITE (3,20) C
20 FORMAT (I 5)
   IF (C-1000) 30,40,40
30 A=B
   B=C
   GO TO 10
40 STOP
   END
```

2. Null operators are often called delimiters provided they do not cause the generation of object code. List the delimiters in the FORTRAN program of Exercise 1.

*3. Revise the algorithm in Figure 5-5 so as to accommodate the multi-character variables which were provided by the syntactic changes made in Exercise 4 at the end of Chapter 4. Write out approximately 100 lines of code for your revised table function.

*This exercise is more on the order of a project and will require a great deal of time and effort. The reader may wish to attempt the solution after the rest of the text has been completed.

Chapter 6

The Parse Function

There are many computer scientists who would claim that the parse function, or what might also be called the syntax analyzer, is the most important part of any compiler. No doubt the reason such claims are made is because high-level languages allow many different kinds of statements and, as a result, it takes a lot of checking and sometimes a lot of time to determine whether a statement is legitimate. The literature in this area is overwhelming to the beginner, and differences of opinion as to how one should parse a statement are equally overwhelming. We, therefore, will not attempt to sort through all the available material, but rather, will move as directly as possible to the parsing technique which will handle the PL/W syntax.

First, however, we must introduce the reader to the subject. Consider the simple syntax:

$$\begin{aligned} \langle E\rangle &::= \langle T\rangle | \langle E\rangle + \langle T\rangle \\ \langle T\rangle &::= \langle F\rangle | \langle T\rangle * \langle F\rangle \\ \langle F\rangle &::= i \end{aligned} \qquad (1)$$

This syntax will produce statements such as:

$$i*i+i$$
$$i+i*i$$
$$\vdots$$

Computer scientists have made a distinction between a bottom-up parse and a top-down parse. A bottom-up parse consists of starting with the statement and arriving at the so-called prime symbol which in this case is $\langle E\rangle$. A top-down parse begins with the prime symbol and ends with the statement. A parsing diagram for $i*i+i$ is

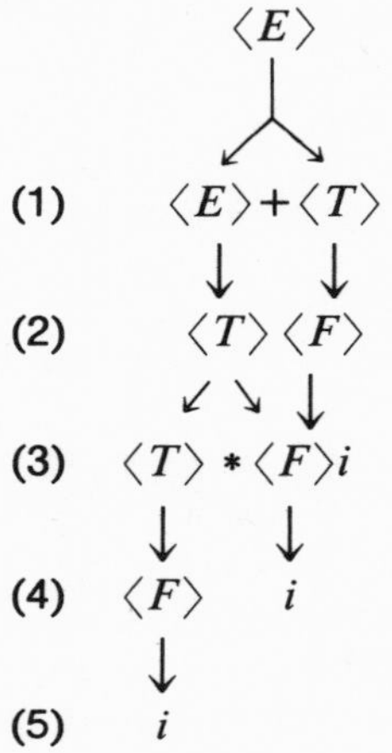

$$
\begin{array}{llcl}
 & & \langle E\rangle & \\
 & & \swarrow\ \searrow & \\
(1) & & \langle E\rangle + \langle T\rangle & \\
 & & \downarrow \qquad \downarrow & \\
(2) & & \langle T\rangle \ \langle F\rangle & \\
 & & \swarrow\ \searrow\ \downarrow & \qquad (2)\\
(3) & \langle T\rangle * \langle F\rangle i & & \\
 & \downarrow \qquad \downarrow & & \\
(4) & \langle F\rangle \qquad i & & \\
 & \downarrow & & \\
(5) & i & &
\end{array}
$$

The numbers at the left are supplied simply for reference purposes. We shall refer to them as levels. Let us assume that the above diagram represents a top-down analysis. Notice that at level (1) we chose $\langle E\rangle+\langle T\rangle$ as the definition for the prime symbol $\langle E\rangle$. This is the second alternative supplied in the syntax and we selected it because we could "see" that it would "work." One of the great difficulties with a top-down syntax analyzer involves making the computer "see" the proper choice. Let us illustrate this difficulty by performing the top-down parse using the alternatives as they appear from left to right in the syntax. We obtain the following:

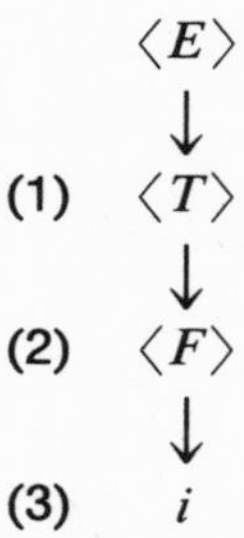

$$
\begin{array}{ll}
 & \langle E\rangle \\
 & \downarrow \\
(1) & \langle T\rangle \\
 & \downarrow \\
(2) & \langle F\rangle \\
 & \downarrow \\
(3) & i
\end{array}
\qquad (3)
$$

Clearly, we end up with only i whereas the statement we are analyzing consists of four more symbols, namely, $*i+i$. The result simply means that the syntax analysis took a wrong turn somewhere. We do not have any other alternatives at level (3) so we could back up to level (2) and obtain the following:

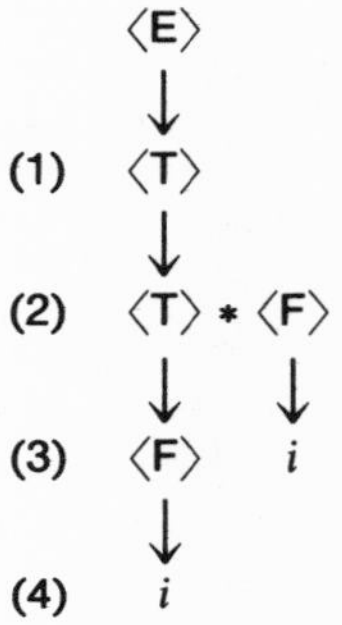

$$\begin{array}{lccc} & \langle E\rangle & & \\ & \downarrow & & \\ (1) & \langle T\rangle & & \\ & \downarrow & & \\ (2) & \langle T\rangle & * & \langle F\rangle \\ & \downarrow & & \downarrow \\ (3) & \langle F\rangle & & i \\ & \downarrow & & \\ (4) & i & & \end{array} \qquad (4)$$

Even though the result here is different from that above, it nevertheless represents another failure to parse the original statement. Not only does it represent that failure, it sets the stage for an interesting feature of the grammar in question.

Suppose we back up once more. Now it is level (3) where the first alternative occurs. We obtain:

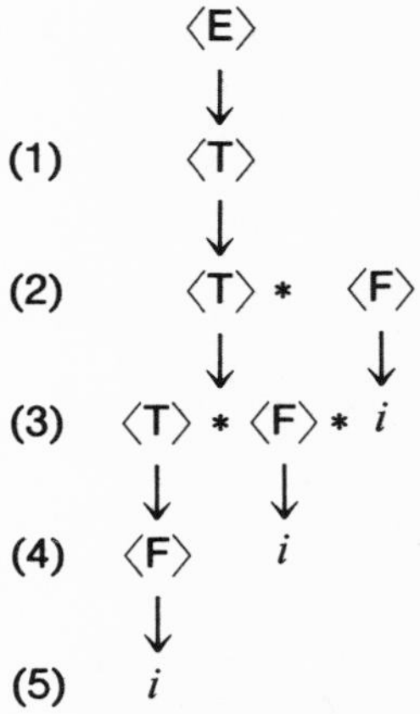

$$\begin{array}{lccccc} & & & \langle E\rangle & & \\ & & & \downarrow & & \\ (1) & & & \langle T\rangle & & \\ & & & \downarrow & & \\ (2) & & & \langle T\rangle & * & \langle F\rangle \\ & & & \downarrow & & \downarrow \\ (3) & \langle T\rangle & * & \langle F\rangle & * & i \\ & \downarrow & & \downarrow & & \\ (4) & \langle F\rangle & & i & & \\ & \downarrow & & & & \\ (5) & i & & & & \end{array} \qquad (5)$$

We have produced the statement $i*i*i$, which is still incorrect and further back up will merely generate an indefinitely long statement of the form:

$$i*i*\ldots*i*i$$

The problem we have illustrated is due to the fact that the grammar in **(1)** uses left recursion; that is, a nonterminal symbol on the left of ::=

occurs again as the leftmost symbol of an alternative on the right of ::= . Left recursion is only one example of a difficulty that is due to some feature of the grammar being used. And so the question becomes one of designing a grammar which is either free of such difficulties, or designing an analyzer which can deal with them efficiently. We shall eventually pursue the first of these two alternatives, but before we do so, there are a few remaining introductory comments that should be made.

The problem just illustrated involved left recursion in a top-down parse. The reader should not assume that difficulties that arise in top-down parsing will necessarily arise in bottom-up parsing. The reader should also be aware that removing difficulties such as left recursion from a grammar may at the same time remove some highly desirable statements from the language. Due to numerous attempts to deal with such complications, a large number of grammatical forms have been invented and classified. We have provided a partial list of such classifications so that the interested reader may have an index reference for investigating them elsewhere.

1. Bounded-context
2. Context-free
3. Context-sensitive
4. ϵ-free
5. Left-bounded-context
6. Left-right-bounded-context
7. LR(k)
8. Phrase structure
9. Precedence
10. PR(k)
11. Regular
12. Right-bounded-context

6-1 Regular Grammars

The grammatical structure that will interest us is the so-called regular grammar. It has some desirable features and allows a large variety of statements. A very useful device for generating such a grammar is to graph on a network of nodes and arrows the statements desired. We begin with a starting node:

Using the statements generated by the grammar in (1) above we see that all such statements must begin with an "*i*." This is graphed as follows:

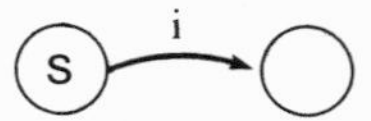

The "*i*" must be followed by "*" or "+." We therefore obtain:

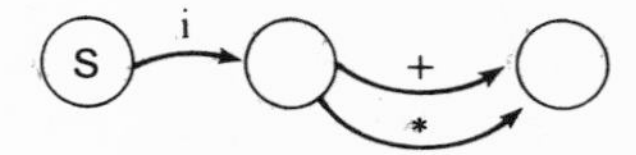

The "*" or "+" is always followed by an *i* which could, once again, be followed by either "*" or "+." This can be graphed as follows:

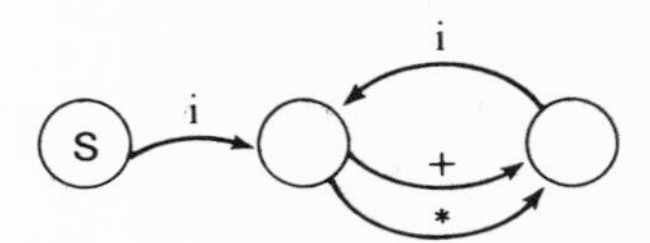

And finally, we have a terminal node.

(6)

The concept behind the graph in **(6)** is that if one always begins at S and ends at T, the symbol string generated will be a legitimate statement in the grammar. If a syntax is regular, then it possesses such a graph. Usually, however, it is much easier to begin with the graph and generate the regular syntax rather than the other way around.

Some relatively simple rules have been invented which allow one to convert the graph of a regular grammar to its syntactic representation. These rules assume a label for every node in the graph. Consequently, we will supply such labels in **(6)** as follows:

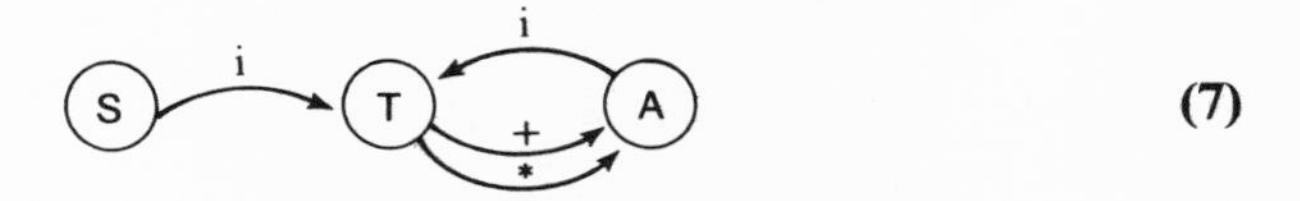

(7)

An arrow from the starting node S to any other node, say Q, appears in the graph as

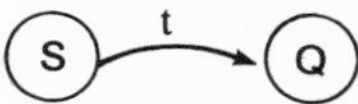

where t is some terminal symbol. In the syntax this becomes

$$\langle Q \rangle ::= t$$

An arrow from any node other than S, say P, to any other node, say Q, appears in the graph as

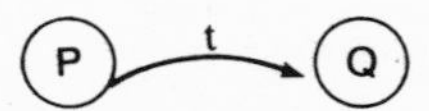

and can be expressed syntactically by

$$\langle Q \rangle ::= \langle P \rangle t$$

Application of these rules to (7) yields the following:

$$\begin{aligned} &\langle T \rangle ::= i \\ &\langle A \rangle ::= \langle T \rangle + \\ &\langle A \rangle ::= \langle T \rangle * \\ &\langle T \rangle ::= \langle A \rangle i \end{aligned}$$

Combining the definitions of A and T we obtain:

$$\begin{aligned} &\langle T \rangle ::= i | \langle A \rangle i \\ &\langle A \rangle ::= \langle T \rangle + | \langle T \rangle * \end{aligned} \qquad \textbf{(8)}$$

The syntax here is free of recursion but it still must back up when performing a top-down parse. It will, however, perform a bottom-up parse with a great deal of "regularity" as is indicated by the examples below:

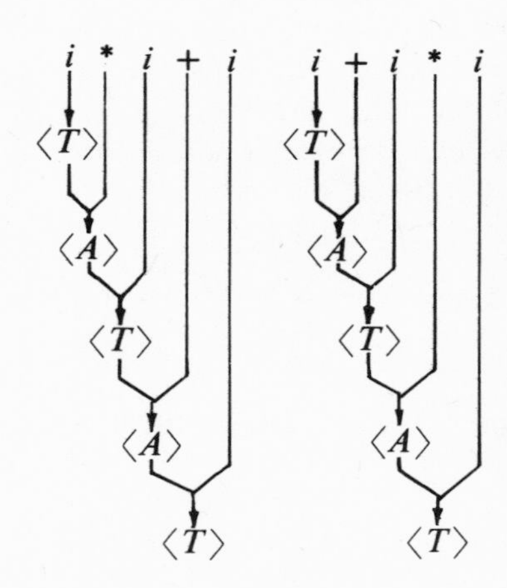

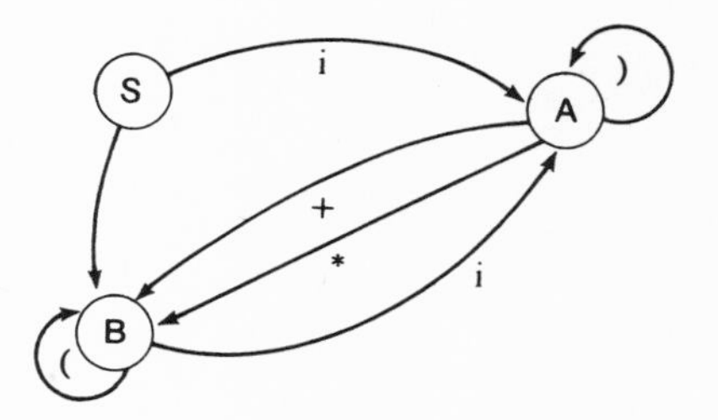

FIGURE 6-1 Directed Graph for Simple Arithmetic Expressions Containing Parentheses [*Note*: A is the terminal node]

In order that the reader not become too enthused about regular grammars we shall point out one of the problems that arises in the implementation of parenthetical expressions. Figure 6-1 displays the graph of the grammar in **(9)** which is merely the grammar in **(1)** with parenthetical expressions added.

$$\begin{aligned} \langle E \rangle &::= \langle T \rangle | \langle E \rangle + \langle T \rangle \\ \langle T \rangle &::= \langle F \rangle | \langle T \rangle * \langle F \rangle \\ \langle F \rangle &::= (\langle E \rangle) | i \end{aligned} \tag{9}$$

The syntax for Figure 6-1 is as follows:

$$\begin{aligned} \langle A \rangle &::= i | \langle B \rangle i | \langle A \rangle) \\ \langle B \rangle &::= (| \langle A \rangle + | \langle A \rangle - | \langle B \rangle (\end{aligned} \tag{10}$$

The reader will note that not only has left recursion occurred once again, but also the syntax will accept symbol strings with unmatched parentheses. Consequently, a separate check must be made to insure that the number of left parentheses is equal to the number of right parentheses.

6-2 A Regular Grammar For PL/W

As was pointed out earlier, it is easier to begin with the graph and then generate the regular syntax. And, as we shall soon see, it is often possible to design the parse function directly from the graph.

In the case of PL/W we will follow a few simple steps. First, we have already put together a syntactic definition of a PL/W program without considering its grammatical features. This is the syntax in Table 3-1.

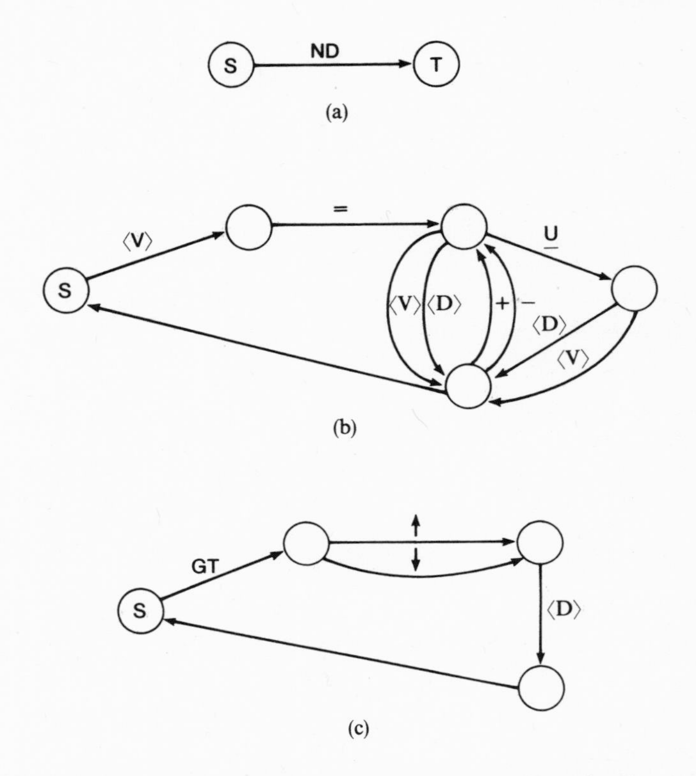

FIGURE 6-2 Directed Graphs for (a) the End Statement; (b) the Assignment Statement; (c) the Transfer Statement (continued on following page)

Secondly, we will construct a graph for each type of statement in a PL/W program. There are five such statements and their graphs are provided in Figure 6-2. Due to spatial considerations we have used abbreviations for the symbols as noted in Figure 6-2.

These graphs can be combined as shown in Figure 6-3.

Using the rules supplied in section 1 of this chapter, we could now begin to generate the regular syntax from Figure 6-3. This is quite a tedious task and as an exercise the reader may wish to write out the regular syntax for some portion of Figure 6-3, for example, the portion defining the assignment statement.

An exciting feature of the graph of a regular grammar is that it is often easier to design the parse function directly from the graph and bypass the generation of the regular syntax altogether.

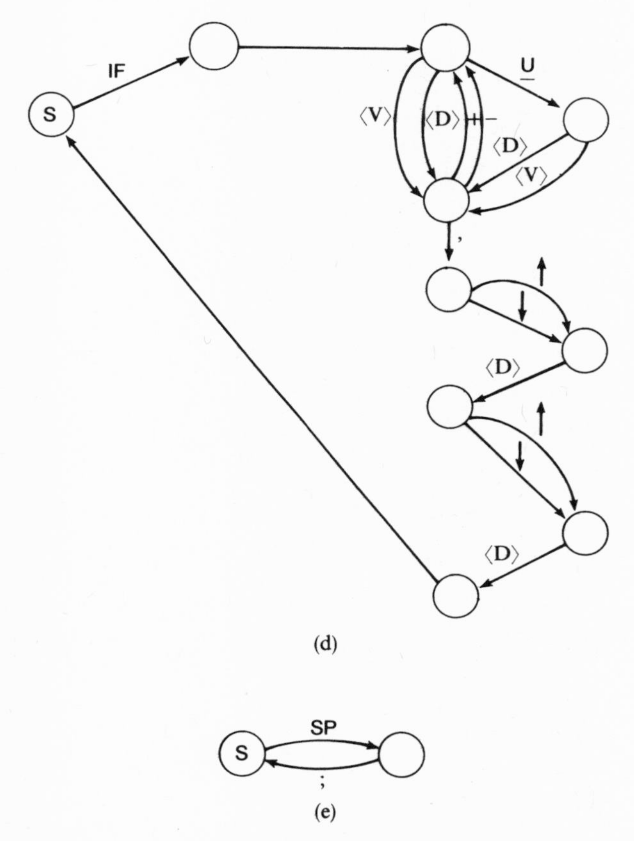

FIGURE 6-2 (d) the Conditional Statement; (e) Stop Statement

6-3 The PL/W Parse Function

The parse function for the PL/W syntax is best approached by first constructing a somewhat generalized flow chart and then adding the necessary detail in the coding procedure. Also, it would be more tedious than instructive to continue our practice of writing the actual code. Therefore, we will be concerned only with the development of a detailed flow chart for the portion of the parse function that analyzes the assignment statement (Figure 6-2(b)). The generalized flow chart for the assignment statement is provided in Figure 6-4. The reader should note that a

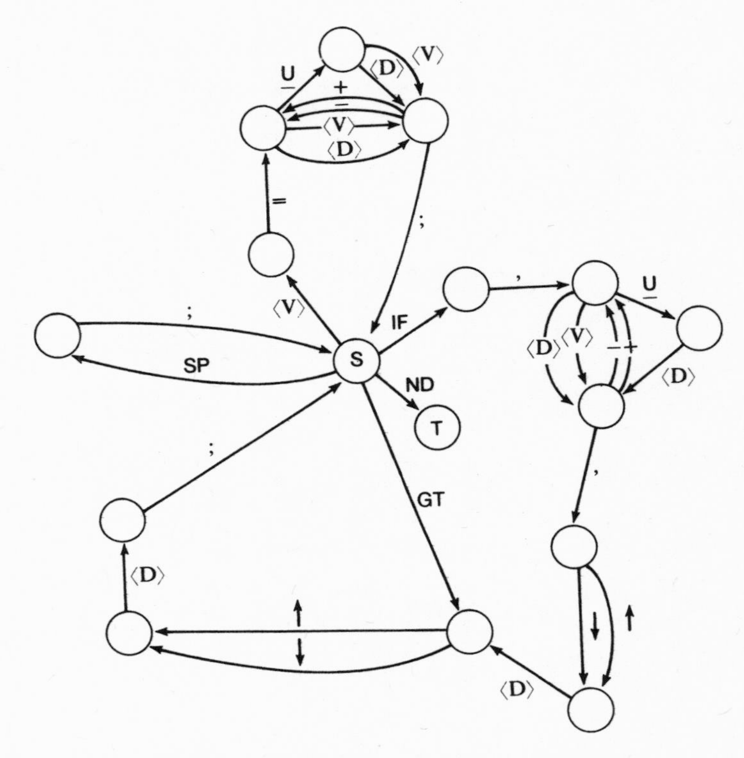

FIGURE 6-3 Directed Graph for PL/W Syntax

diamond with a terminal or nonterminal inside it should be interpreted by the question, "Is the current symbol equal to this particular terminal or a terminal in the set symbolized by this particular nonterminal?"

Several comments are in order regarding Figure 6-4. First, the process for getting the first or next symbol is merely a matter of advancing a pointer by one or two. For the pointer we will use BS which must be initialized to point to the beginning of the symbol string. We can then check for the marker. Assuming the marker is incurred, we may advance the value of BS by one to obtain the first symbol. Because the PL/W syntax allows only single-character variables in an assignment statement, we can advance BS by two to obtain the rest of the symbols. This will have the effect of skipping over the marker.

There is a problem worth mentioning which arises when testing $\overline{BS}$ for equality with one of the IC codes, for example, $\overline{BS}=22$. A single subtraction such as $\overline{BS}$-22 is not sufficient. If $\overline{BS}$-22 is negative, we know $\overline{BS}$ is not equal to 22, but, if $\overline{BS}$-22 is positive, we do not know that $\overline{BS}$ is equal to 22

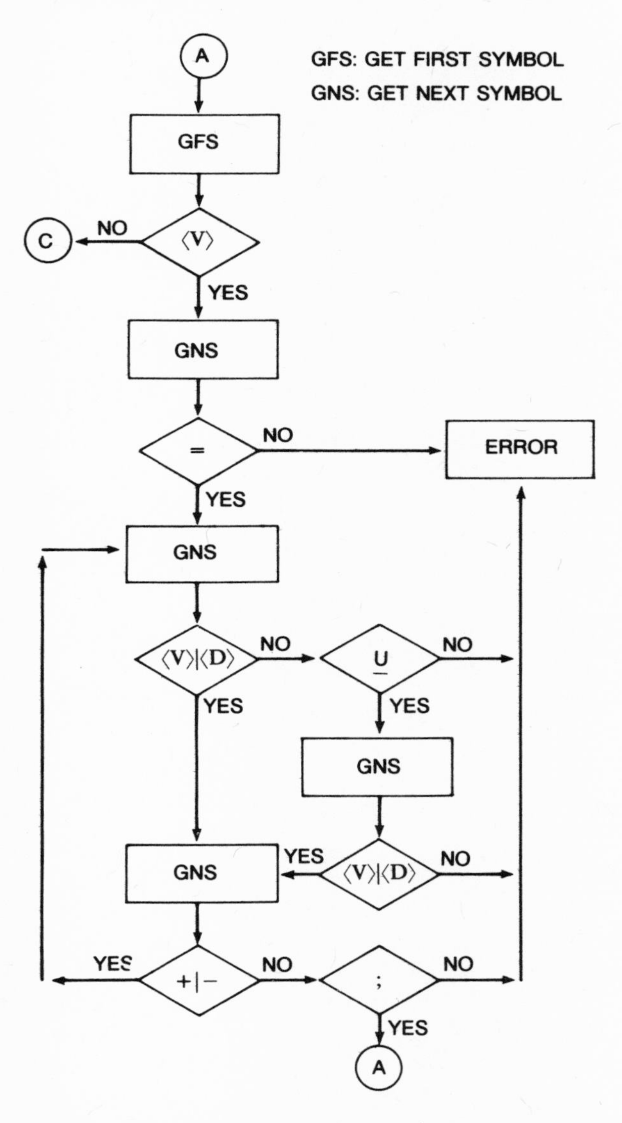

FIGURE 6-4 General Flow Chart for PL/W Assignment Statement

because it could be greater than 22. Therefore, a second subtraction of the form 22-$\overline{BS}$ is required. If both subtractions yield positive results, then $\overline{BS}=22$.

In Figure 6-4 we have an error block. This is something new and as a result we must specify the results of the error function. For sake of brevity we will implement an extremely simple error function while reminding the reader that error functions can be very extensive. For example, in compilers for some of the high-level languages there are diagnostics provided for the user. These diagnostics are used to guide the programmer in the task of debugging the program. Some compilers provide simple diagnostics that may only inform the reader of a syntax error, whereas others may classify the error, locate by line and column, and suggest corrective measures. In our case we will fill the accumulator with all ones and stop.

This is not a foolproof error function. There may be several times during the execution of a program when the accumulator contains all ones. If one of these times represents the final result for a particular program, then the execution of that program will terminate with an error condition displayed when in fact no error has occurred.

One remedy for the above problem can be found in the way the programmer tests his or her program. Consider a program in a high-level language such as FORTRAN. One can imagine running a program that is free of syntax errors, end-of-file errors, and other error types, and yet is not executing properly. Perhaps a minus sign should have been a plus. A programmer detects such an error by running a test case where the final results are known. Such test cases must be carefully constructed so as to guarantee that the test case checks every executable statement in the program. They must also be constructed so as to insure that an improperly coded statement will not, by coincidence, give the correct result.

Suppose a program contained the statement A=B*C and the programmer miscoded the statment as A=B+C. Both statements could be syntactically correct and if the test case set B=2 and C=2, the final result, A=4, would be obtained. Clearly, B=2 and C=2 is a poor test case for this statement.

Therefore, a test case for a PL/W program must be designed in such a way as to produce a final result that will not place all ones in the accumulator.

The error function itself can be coded as follows:

E_1:CLA
E_2:ADD "777"
E_3:STP

The locations E_1, E_2, and E_3 can be placed at the end of the compiler coding.

The tests for variables and integers take advantage of the structure of the IC codes in Table 4-1. Any time an IC code is incurred which is less than 22 we know we have either a variable or an integer. This allows for relatively simple coding of these tests.

Before we supply the detailed flow chart for Figure 6-4, we should explore one final aspect of the PL/W parse function. When the first symbol is tested we transfer to C in the event that it is not a variable. The reader may assume that C is the beginning of that portion of the parse function which tests the conditional statement. These statements begin with either GT or IF. $\overline{\text{BS}}$ at this point will be the second character of what is presumably a multi-character symbol. We cannot be content to check for either the T of GT or the F of IF without first backing up BS and checking for the G or the I. Neglecting to do so would mean that our parse function would accept such things as NT for GT and PF for IF. Later on in the stack function when the compiler attempts to generate code for these operators, it would not be able to locate them in the operator type table.

Figure 6-5 is the detailed flow chart for Figure 6-4. The reader will note that a great deal of PREP coding will be required because the compiler variable $\overline{\text{BS}}$ is examined in every test. However, there is no operation which cannot be accomplished with the eight operations supplied in the WHYCO hardware.

The ambitious reader is encouraged to explore the effects of allowing multi-character variable names and multi-character integers in the PL/W syntax. A thorough grasp of Figure 6-5 and its coded implementation will provide a good background for these explorations. Things become much more complex when these extensions of the syntax are allowed and the reader should study them in at least a moderate amount of detail before turning to the more advanced texts on compiler design and implementation.

Figure 6-5 represents the portion of the parse function that analyzes the PL/W assignment statement. It should not be difficult to develop the rest of the parse function in a similar manner. For example, at (C) we can check for an IF, and if successful, follow it with a test for the comma. If we do not incur the comma, then we have an error. If we do, we begin checking for a PL/W expression which is nearly the same as the lower right-hand portion of Figure 6-5, the main difference being the check for $\overline{\text{BS}}=23$ (;). By referring to Figure 6-2(d) the reader can see that the check should be $\overline{\text{BS}}=22$ (,). If the comma is missing, we have an error. If not, we continue the flow by checking for the directionals and integers which

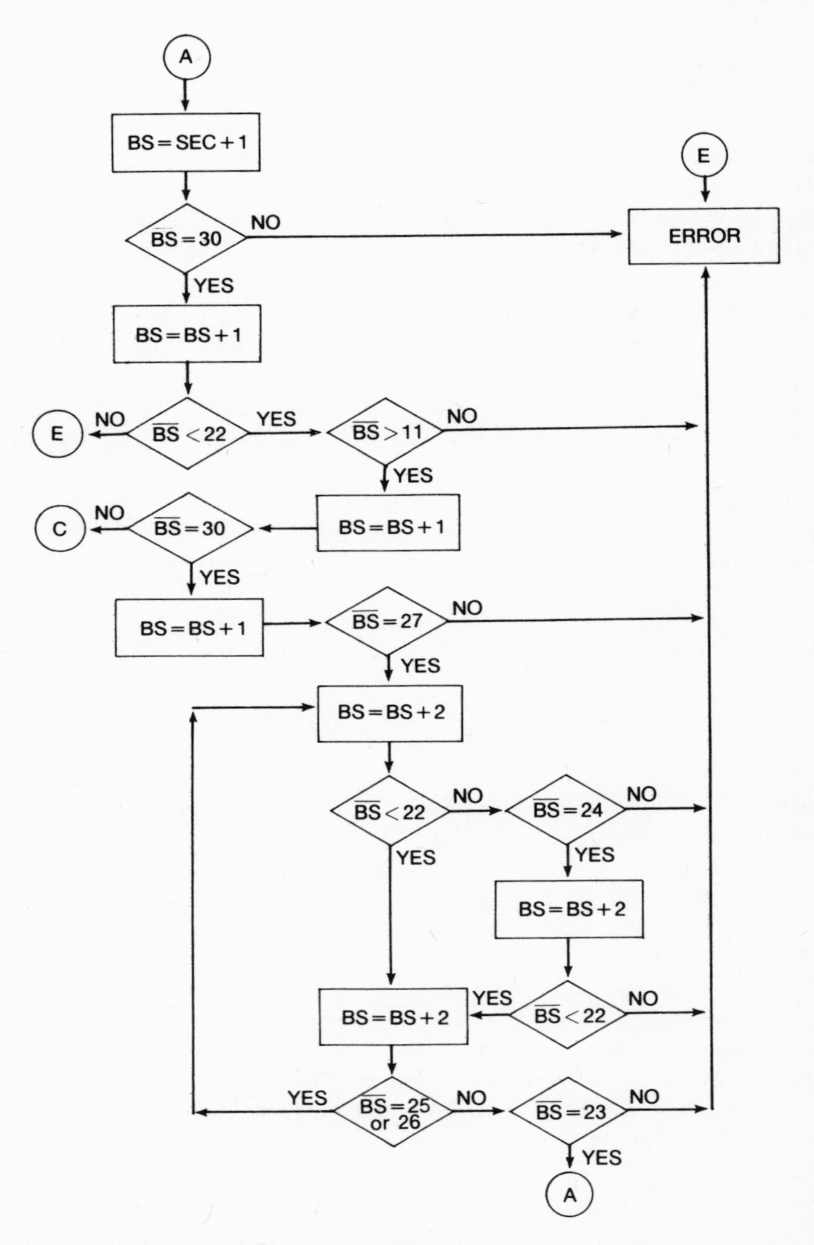

FIGURE 6-5 Algorithm for Parse of PL/W Assignment Statement

conclude the PL/W IF statement. The remaining parts of the parse function are quite simple.

At this point the reader may wish to give some serious consideration to exercise 2.

EXERCISES

1. Using the rules supplied in section 6-1, write out the regular grammar for Figure 6-3.
2. Design a flow chart similar to Figure 6-5 which will parse the PL/W IF statement.
3. Modify Figure 6-2(d) in light of the syntactic change suggested in the statement on p. 107. Draw the flow chart for your modified version of Figure 6-2(d).
4. Using the PL/W syntax as provided in Table 3-1, draw the parsing diagram for the program on p. 44.

Chapter 7

The Stack Function

When the parse function completes its task we will have accomplished one of two things. Either we will have incurred an error in the coding, in which case we would have incurred the STP code at E_3 in the error function, or we will enter the stack function with the assurance that the coding is syntactically correct.

The stack function rearranges each statement into a form which is more convenient for the generation of the object code. The rearrangement we have in mind is referred to as postfix notation.

7-1 Postfix Notation

The reader is familiar with simple algebraic equations, and because of this, we will use them to develop the concepts necessary for the design of the stack function. Let us consider first some binary operators such as $+$, $-$, $*$, $\div$, and $=$. Such operators are referred to as "binary" because they require two operands. There are operators other than binary operators (see Table 5-5), and we will eventually consider them as well.

A number of graphic devices have been developed which make it relatively easy to convert expressions into postfix notation. These expressions occur in a form that computer scientists have come to call infix notation. It simply means that the operator occurs *in* between the operands. The reader may have guessed that postfix means that the operator occurs after its operands, and prefix means the operator occurs before the operands. Table 7-1 contains some infix expressions and their corresponding postfix forms.

Notice how easy it is to comprehend the changes in (1) and how difficult it is to comprehend the changes in (3). The graphic devices alluded to above are most helpful in describing the techniques for converting infix to postfix. One of these graphic devices is the "tree." We might point out in passing that graphic devices can also be implemented as data structures in a computer program. This is especially easy to do in some of the high-level

Table 7-1
Infix and Postfix Strings

	INFIX	POSTFIX
(1)	N=9	N9=
(2)	F=P+T	FPT+=
(3)	A=((B+C)*D)÷(E*(F+G))	ABC+D*EFG+*+=

programming languages. The infix expressions in Table 7-1 are presented below together with their corresponding trees.

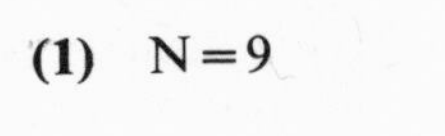

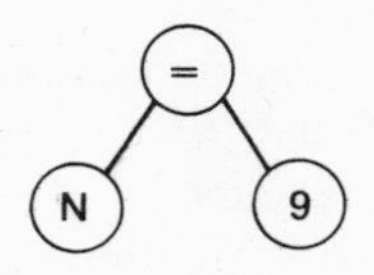

(2) F=P+T

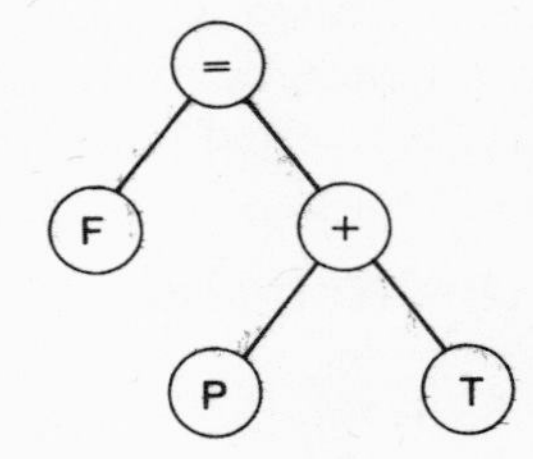

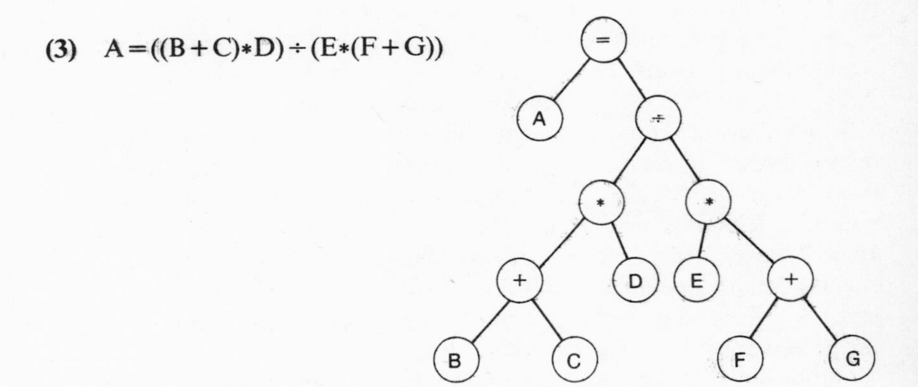

We shall begin our study of infix to postfix conversion by developing an algorithm which will produce the postfix string from the tree structure. To do this we must label each node in the tree as either an operator node or a terminal node. Let us use the tree in **(3)** above as an example. This tree appears in Figure 7-1.

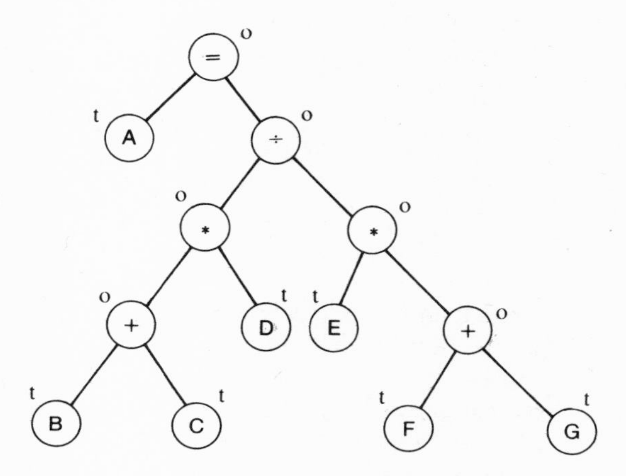

FIGURE 7-1 Tree Representation of $A=((B+C)*D)\div(E*F+G))$

The algorithm requires that we define several parts of the tree. These parts have special names and are listed and defined as follows:

1. Left-most terminal node: defined as the node which one arrives at by starting at the top of the tree and following left branches until arriving at a terminal. If a node that has only a right branch is incurred, then follow that branch and continue as before.
2. Branch: defined as any line connected to a node at its lower end.
3. Simple sub-tree: defined as a node with either a left branch or a right branch or both.

The above definitions are illustrated below.

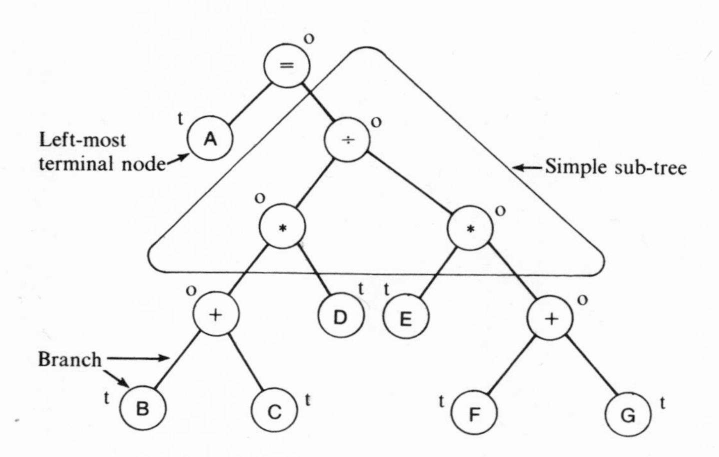

7-2 Infix to Postfix Conversion

With these definitions we can describe an algorithm which will convert the infix expression represented by the tree to postfix.

Step 1: Locate the left-most terminal node. Call this node N. IF none exists, THEN stop.

Step 2: Locate the simple sub-tree containing N as one of its nodes. Call this simple subtree T.

Step 3: Record N and then delete the left branch containing N from the tree. If T has no left branch, go to step 4.

Step 4: IF: the right branch of T contains a terminal node, THEN: (1) record that terminal node, (2) delete the right branch of T from the tree, and (3) change the status of the remaining node of T from an operator node to a terminal node, (4) return to step 1. ELSE: return to step 1.

We have applied the algorithm in the above list to the tree in Figure 7-1 in a step-by-step manner. The application in Figure 7-2 shows only two passes through the algorithm. When the application is completed, which the reader may do as an exercise, the final result will be the postfix form of the original infix expression.

The trees in Figure 7-2 are called binary trees for obvious reasons. How can a binary tree be used to "graph" an expression that uses nonbinary operators? Using "null" operators and "null" terminals, both of which will be symbolized by $\perp$, makes it possible to construct binary trees for expressions containing nonbinary operators. Figure 7-3 contains two exam-

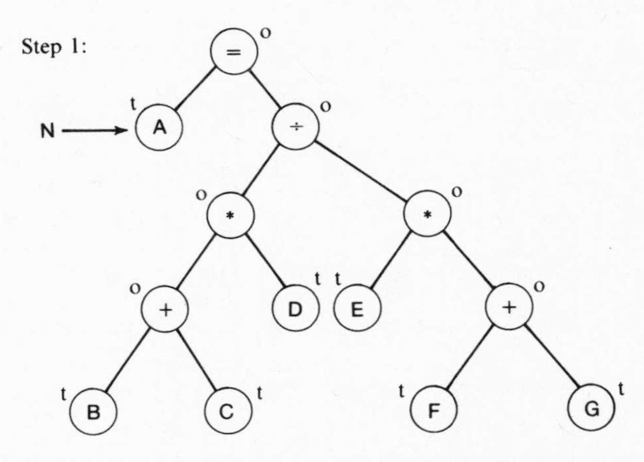

FIGURE 7-2 Infix to Postfix Conversion Using Tree Representation

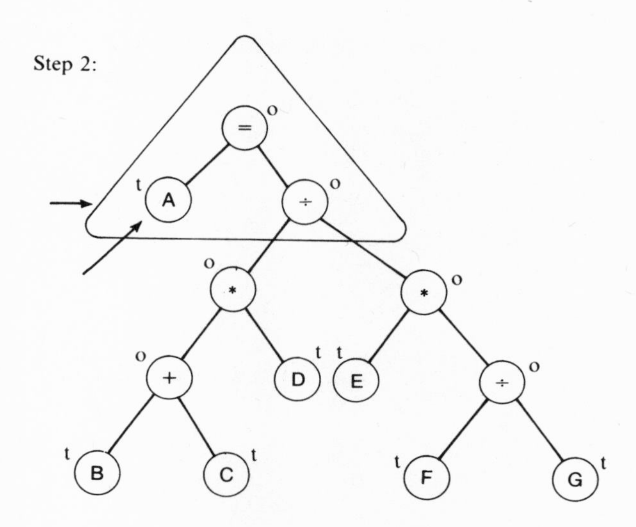

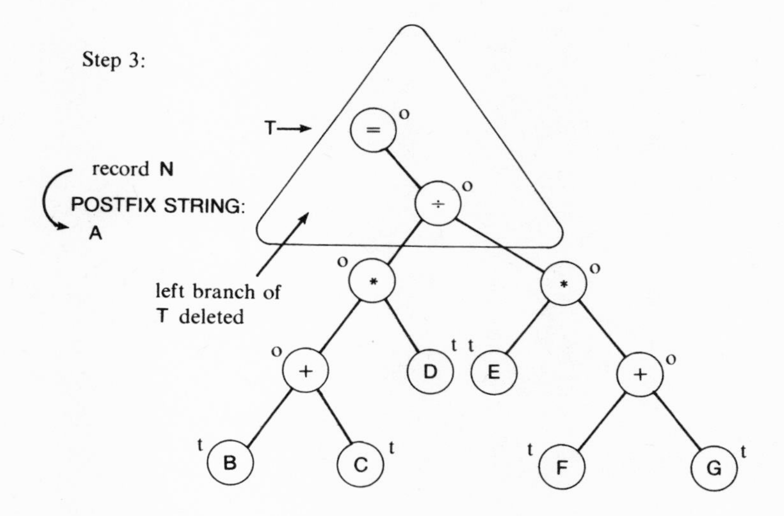

Step 4: Since the right branch of T does not contain a terminal node, ÷ being an operator node, we return to step 1.

FIGURE 7-2 (Continued)

Step 1:

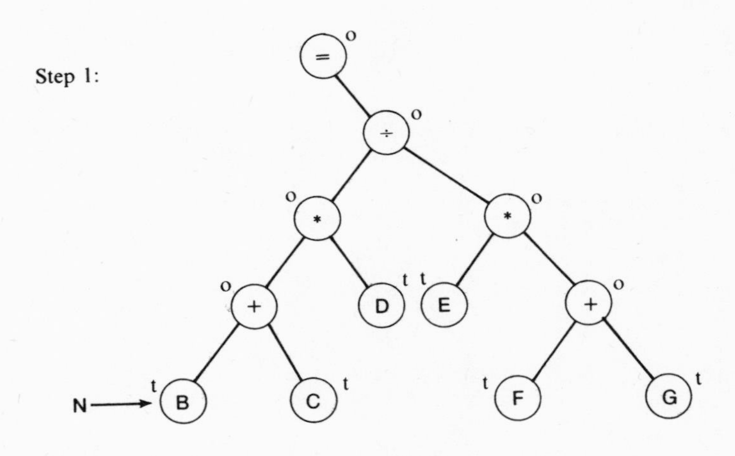

Step 2:

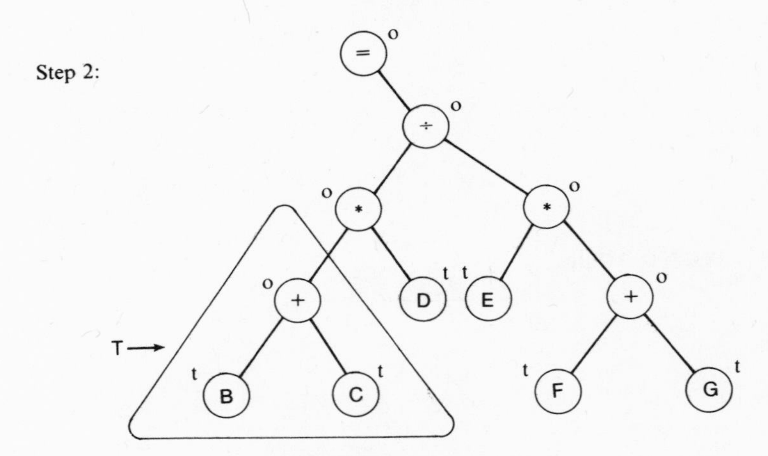

FIGURE 7-2 (Continued)

ples from our example program on the Fibonacci sequence. The ⊥ is ignored in the final postfix string.

The question that now arises concerns the implementation of the postfix to infix conversion. If we were writing a compiler in a high-level language such as PL/1, it would be relatively easy to implement the algorithm described in the list on p. 94 and displayed in Figure 7-2. Our task, however, is to implement the infix to postfix conversion in machine code, using only the eight operations of the WHYCO.

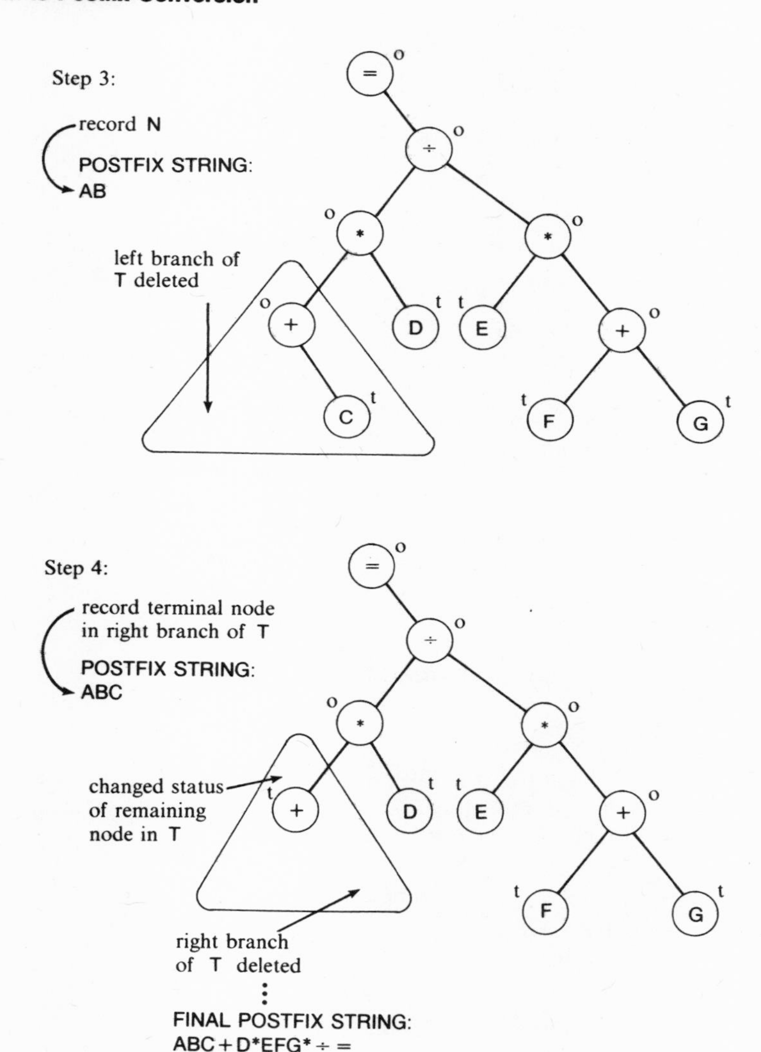

FIGURE 7-2 (Continued)

In passing, we might make a comment or two about compilers which are written in high-level languages. Remember that any compiler written in a high-level language is simply a program, which, like any other program, must be compiled. The result, a "compiled compiler" may then be used to compile programs written in the high-level language for which the original high-level language compiler was written. This process can be described

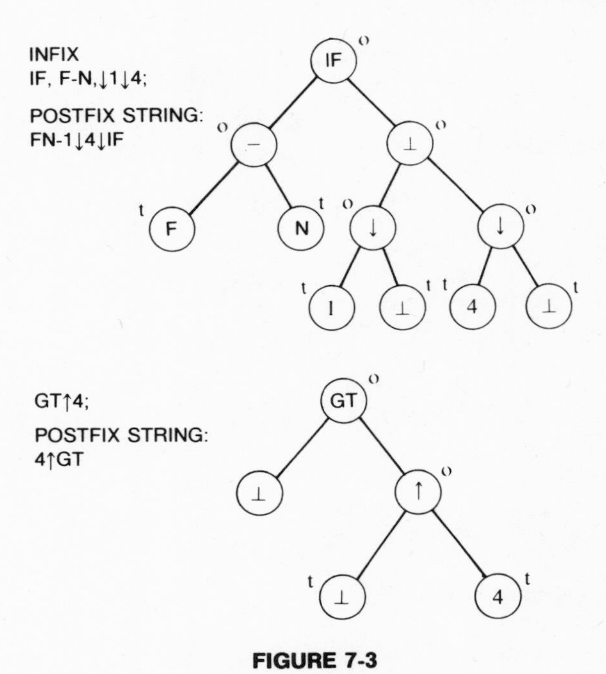

FIGURE 7-3

symbolically as follows. Let:

$A(B)\rightarrow C$: Mean A compiles B into C
H: High-level language
L: High-level language other than H
MC_x: Machine code for x
HC_x: High-level language compiler for x
MCC_x: Machine code compiler for x
P_x: Program written in x

Suppose you write a compiler for L in H and use it to process P_L. The following would take place:

$$MCC_H(HC_L(P_L))\rightarrow MCC_L(P_L)\rightarrow MC_{P_L}$$

We are concerned with MCC_L where L is PL/W, that is, we are concerned with:

$$MCC\text{PL/W}(P\text{PL/W})\rightarrow MC_P\text{PL/W}$$

Before returning to the problem of converting infix to postfix for PL/W, let us first supply some motivation for using the postfix form. The idea is to read the postfix string from right to left. When we incur an operator we place it in a structure called a push down stack. We also record its type, which can be obtained from the operator type table (see section 5-2). The same thing is done when an operand is incurred. It is placed on a different push down stack, and, at the same time, the type number of the operator on the top of the operator push down stack is reduced by one. When an operator on the top of the operator push down stack has a type number equal to zero we can generate code for that operation. For the moment, we will indicate code by a box around the coded expression. The operator is then deleted from the top of the operator stack and the coded expression is treated as a single operand and returned to the top of the operand stack. When it is returned the type number of the "new top" of the operator stack is reduced by one. As long as the type number of the top operator is not zero, we will continue to "place" on the stacks the symbols from the postfix string. The process is complete when the operator stack is empty. Table 7-2 displays the process for the postfix string in (**3**) on p. 92. The "left" of the stack is the "top," and the "→" indicates that the rightmost symbol of the postfix string is going to be processed.

7-3 THE PL/W STACK FUNCTION

Implementing Table 7-2 in WHYCO machine code could be extremely tedious, especially when we consider the fact that the infix to postfix conversion is much more involved than some, if not all, of the functions of the compiler which we have already studied. The reader will also note that no mention was made of the process by which the input string was converted to the tree structure. This omission is partly due to the fact that the tree structure is primarily, at least as far as the PL/W compiler is concerned, a conceptual tool. That is to say, we will not attempt to implement the tree structure in the compiler for PL/W. Another reason for this omission lies in the fact that the PL/W syntax does not provide for parenthetical expressions. Let us examine the consequences of this aspect of the PL/W syntax for infix to postfix conversion.

Consider the following PL/W infix statement, its tree, and its postfix form:

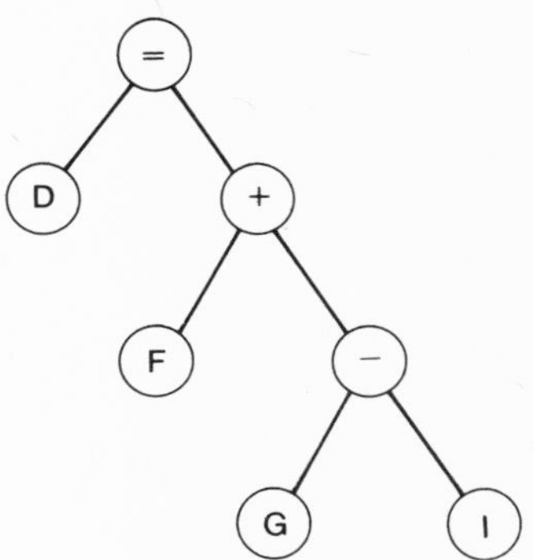

Table 7-2
Stack Processing of Postfix Strings

Postfix String	Operator Stack	Type Number Stack	Operand Stack
ABC+D*EFG+*÷=→	=	2	—
ABC+D*EFG+*÷→	÷=	2,2	—
ABC+D*EFG+*→	*÷=	2,2,2,2	—
ABC+D*EFG+→	+*÷=	2,2,2,	—
ABC+D*EFG→	+*÷=	1,2,2,2	G
ABC+D*EF→	+*÷=	0,2,2,2	FG
ABC+D*E	*÷=	1,2,2	[F+G]
ABC+D*E→	*÷=	0,2,2	E [[F+G]]
ABC+D*	÷=	1,2	[E* [F+G]]
ABC+D*→	*÷=	2,1,2	[E* [F+G]]
ABC+D→	*÷=	1,1,2	D [E* [F+G]]
ABC+→	+*÷=	2,1,1,2	D [E* [F+G]]
ABC→	+*÷=	1,1,1,2	CD [E* [F+G]]
AB→	+*÷=	0,1,1,2	BCD [E* [F+G]]
A	*÷=	0,1,2	[B+C] D [E* [F+G]]
A	÷=	0,2	[[[B+C] *D] ÷ [E* [F+G]]]
A	=	1	[[[B+C] *D] ÷ [E* [F+G]]]
A→	=	0	A [[[B+C] *D] ÷ [E* [F+G]]]
–	–	–	[A= [[[B+C] *D] ÷ [E* [F+G]]]]

It would also be legitimate to treat the same statement in the following manner:

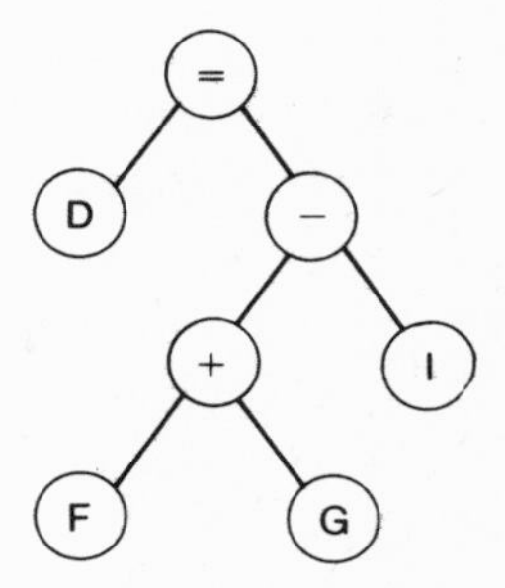

This ambiguity will cause no problem in the PL/W compiler because "+" and "−" have no precedence over one another. This would not be the case if an operator such as "*" were allowed. Consider the following:

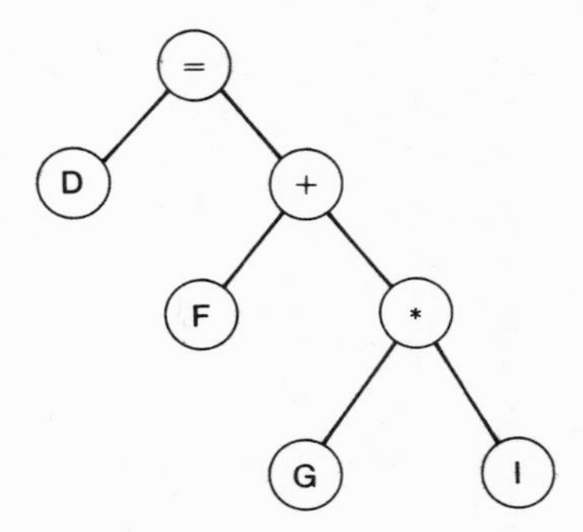

This statement reflects what is normally intended, namely, that multiplication be performed before addition, as occurs in common algebra. Precedence of addition over multiplication is reflected in the following statement:

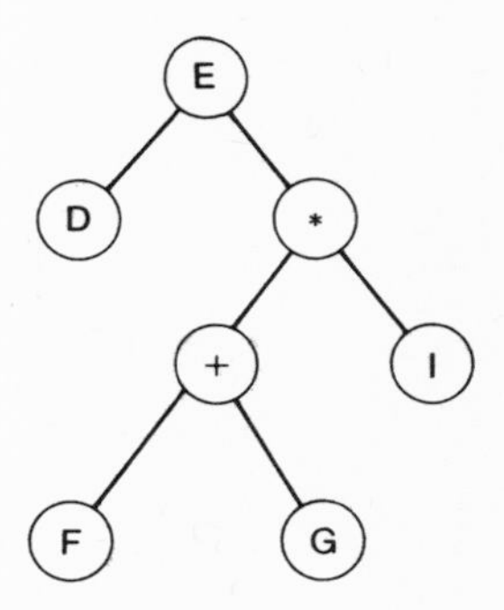

The obvious solution to the problem lies either in the use of parentheses or in developing precedence indicators for the operators. Such indicators would be part of the operator type table. Because the problem we have been discussing will be only partly solved in the PL/W compiler, let us proceed to our technique for converting infix to postfix in the PL/W compiler and then return for a brief treatment of how the problem could be handled when parenthetical expressions are allowed, as they are in most high-level languages.

The trees on pp. 100 and 101 illustrate the lack of precedence among the PL/W operators "+" and "−". This fact will allow us to read code containing "+" and "−" from left to right and place the operands on an operand stack and the operators on an operator stack as they are incurred.

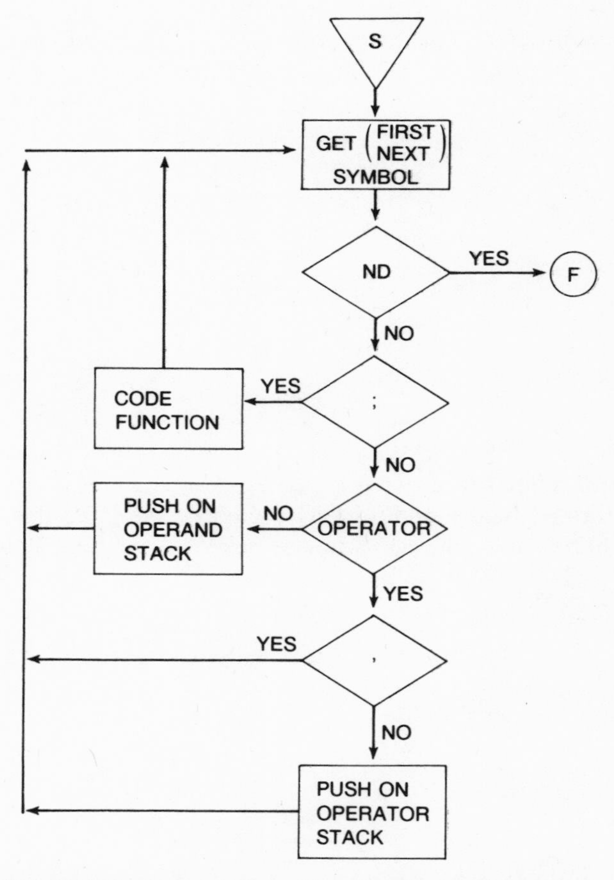

FIGURE 7-4 General Flow Chart for the Stack Function

We may then simply unstack the stacks and generate the code as we go. This is also the technique we would use if we wanted to obtain the postfix string. For the moment let us illustrate the coding and postfix string by drawing a box around the postfix string to indicate a coded expression. This is similar to what we did in Table 7-2, except that we started with the postfix string and drew the box around the infix form of the coded expressions.

The stacking and unstacking will be indicated by the terms PUSH and POP. We "push" a symbol onto a stack and we "pop" it off the stack. A preliminary flow chart for the stacking operation is given in Figure 7-4.

The code function, which will be designed in detail later on, begins by unstacking the results obtained by the algorithm in Figure 7-4. The unstacking is essentially a matter of "popping" the top operator followed by "popping" the top T operands, where T is the type number on top of the type number stack. These values are then used to generate the appropriate code. This process is repeated until the operator stack is empty. In most cases, the results of a code generation must be "pushed" back onto the operand stack as long as the operator stack is not empty. A general flow chart for the unstacking operation is given in Figure 7-5.

We shall now use the algorithm in Figures 7-4 and 7-5 to process the first four lines in the Fibonacci sequence program. The reader may wish to refer to the code on p. 44 for a review of these statements. The input string is given on p. 49. Because of its length we will display only its first three and last three symbols. Table 7-3 shows the results of this processing up to but not including the IF statement.

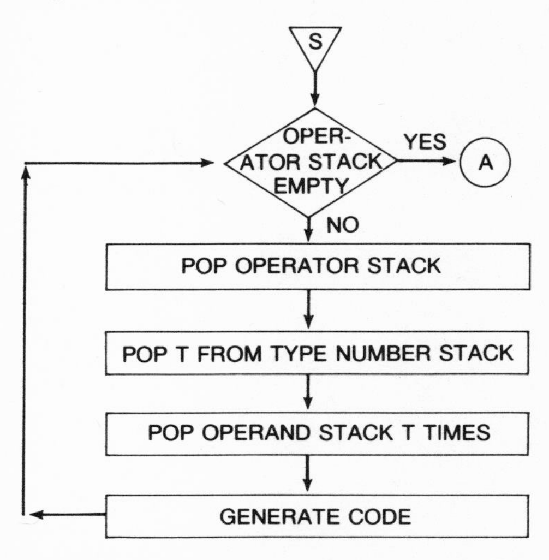

FIGURE 7-5 General Flow Chart for Code Function

Table 7-3

INFIX INPUT STRING	OPERATOR STACK	TYPE NUMBER STACK	OPERAND STACK	STACKCODE
←9; P...;←N=9...; ND	—	—	N	—
←=9;...; ND	=	2	N	—
←9; P...; ND	=	2	9N	—
←; P=...; ND	—	—	—	N9=
←P=1...; ND	—	—	P	—
←=1;...; ND	=	2	P	—
←1; T...;ND	=	2	1P	—
←; T=...; ND	—	—	—	P1=
←T=1...; ND	—	—	T	—
←=1;...; ND	=	2	T	—
←1; F...; ND	=	2	1T	—
←; F=...; ND	—	—	—	T1=
←F=P...; ND	—	—	F	—
←=P+...; ND	=	2	F	—
←P+T...; ND	=	2	PF	—
←+T;...; ND	+=	2,2	PF	—
←T; IF...; ND	+=	2,2	TPF	—
←; IF,...; ND	=	2	PT+ F	—
IF F...; ND	—	—	—	F PT+ =

Let us now consider what would happen if we processed the IF statement in the same manner as the statements which preceded it in the Fibonacci sequence program.

In Figure 7-3 we obtained FN − 1↓4↓IF as the postifx string for the IF statement. The postfix string in the last entry of the code column in Table 7-4 would suggest the following tree:

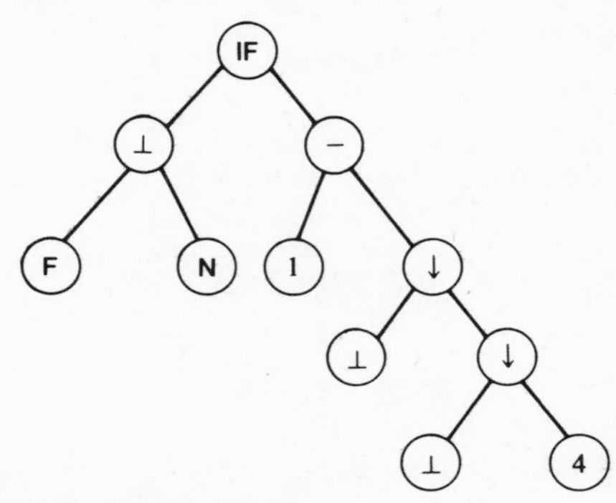

Clearly, this is *not* what is intended.

Table 7-4

INFIX INPUT STRING	OPERATOR STACK	TYPE NUMBER STACK	OPERAND STACK	CODE
←IF,F...;ND	IF	3	–	–
←,F− ...;ND	IF	3	–	–
←F−N...;ND	IF	3	F	–
←−N,...;ND	−IF	2,3	F	–
←N,↓...;ND	−IF	2,3	NF	–
←↓1↓...;ND	↓−IF	1,2,3	NF	–
←1↓4...;ND	↓−IF	1,2,3	1NF	–
←↓4;...;ND	↓↓−IF	1,1,2,3	1NF	–
←4;P...;ND	↓↓−IF	1,1,2,3	41NF	–
←;P= ...;ND	↓−IF	1,2,3	[4↓] 1NF	–
P=T...;ND	−IF	2,3	[[4↓] ↓] 1NF	–
P=T...;ND	IF	3	[1 [[4↓] ↓] −] NF	–
P=T...;ND	–	–	–	[FN 1 [[4↓] ↓] − IF]

The problem lies in the fact that the "↓" and the "−" have precedence over "IF," whereas our stack function treats them as equal in precedence. The solution to the problem is relatively simple. First, in regard to the directionals "↑" and "↓." These pseudo-operators will be used to record transfer information in tables. The use of such tables will be explained in a later section. For the moment, the most important point concerns the fact that the directionals should not generate any object code. Consequently, we will treat them as unary operators and when they are popped off the operator stack we will simply attach them to the top operand as postfix operators and place the result back on the operand stack. We will draw a circle around these results to indicate that even though they are now part of the postfix string, they do not represent any object code. This procedure will be followed each time the top of the infix input string is an operand and the top of the operator stack is a directional. In Figure 7-6 we assume this will occur in the PUSH OPERAND block. Secondly, every time we incur the "," we will compare the top element (T) of the type number stack with a number (K) which counts the number of operands on the operand stack. If $T \leqslant K$, we will generate the code for that operator. We will examine this technique shortly.

Before we do so, let us return to the problem of parenthetical expressions in general. Our solution suggests the more general solution of generating code for whenever a right parenthesis is incurred in the infix input string. Once we start generating code, we would continue "popping" off operators and generating more code until we incur the left parenthesis which would have been placed on the operator stack at the time it was read. Let us illustrate this procedure using the infix string in (3) of Table 7-1 as input.

Table 7-5

INFIX INPUT STRING	OPERATOR STACK	TYPE NUMBER STACK	OPERAND STACK
←A=((B+C)*D÷(E*(F+G))	—	—	A
←=((B+C)*D÷(E*(F+G))	=	2	A
←((B+C)*D)÷(E*(F+G))	(=	0,2	A
←(B+C)*D)÷(E*(F+G))	((=	0,0,2	A
←B+C)*D÷(E*(F+G))	((=	0,0,2	BA
←+C)*D)÷(E*(F+G))	+((=	2,0,0,2	BA
←C)*D)÷(E*(F+G))	+((=	2,0,0,2	CBA
←)*D)÷(E*(F+G))	(=	0,2	[BC+] A
←*D)÷(E*(F+G))	*(=	2,0,2	[BC+] A
←D)÷(E*(F+G))	*(=	2,0,2	D [BC+] A
←)÷(E*(F+G))	=	2	[[BC+] D*] A
⋮	⋮	⋮	⋮

The reader may complete Table 7-5 as an exercise and upon doing so will discover that the final result in the operand stack, which represents a coded expression in postfix form, is the same postfix form generated by the algorithm which operates on the tree structure for the given expression (see Figure 7-2). This concludes the brief treatment of parenthetical expressions alluded to in the discussion on p. 102.

As we have pointed out, the "," in PL/W plays a role similar to "(" and ")" in other high-level languages, that is, it controls precedence of one operator over another. Because of the syntax of PL/W, this precedence occurs only in the IF statement. It would have been easy to avoid this problem altogether by simply redefining the syntax for the IF statement in the following manner:

⟨CONDITIONAL⟩::=

IF, ⟨VARIABLE⟩, ⟨DIRECTIONAL⟩⟨INTEGER⟩⟨DIRECTIONAL⟩⟨INTEGER⟩

Having done so, however, would have sidestepped a very important aspect of the compiling process. It would nevertheless be worthwhile for the reader to consider the consequences of the syntactic change suggested in the above equation.

Let us now illustrate our solution involving the conparison of T and K by applying it to the IF statement in the Fibonacci sequence program.

Table 7-6

Infix Input String	Operator Stack	Type Number Stack (Top = T)	Operand Stack	Value of K	Code
←IF, F−N,↓1↓4;	IF	3	–	0(initial value)	–
←, F−N,↓1↓4;	IF	3	–	0	–
		Since $T \nleq K$ we continue to process the infix input string. Note that we do not place the "," on the operator stack as we did with the "(."			
←F−N,↓1↓4;	IF	3	F	1	–
←−N,↓1↓4;	−IF	2,3	F	1	–
←N,↓1↓4;	−IF	2,3	NF	2	–
←,↓1↓4;	IF	3	[FN−]	1	–
		Here $T \leqslant K$ and therefore code (or the postfix string) was generated for the top operator and its operands. Note that popping off N and F would have reduced K to zero. Also, returning [FN−] to the operand stack would have incremented K to one.			
←↓1↓4;	↓IF	1,3	[FN−]	1	–
←1↓4;	IF	3	(1↓) [FN−]	2	–
←↓4;	↓IF	1,3	(1↓) [FN−]	2	–
←4;	IF	3	(4↓) (1↓) [FN−]	3	–
←;	–	–	–	–	[[FN] (1↓) (4↓) IF]

And finally, the coded postfix expression in the last entry of Table 7-6 is as it should be, that is, its tree structure agrees with (1) of Figure 7-3.

The general flow chart for the stack function which was given in Figure 7-4 will have to be changed to reflect our solution to the precedence problem which occurs in the PL/W IF statement. This change is provided in Figure 7-6.

The actual implementation of Figure 7-6 will require a number of initializations which are not explicit in the flow chart. Table 7-7 contains the variable name, its meaning, its initial value, and an explanatory note.

Let us now consider what must occur when the stack function attempts to get the first or next symbol. BS will be pointing to the first location of

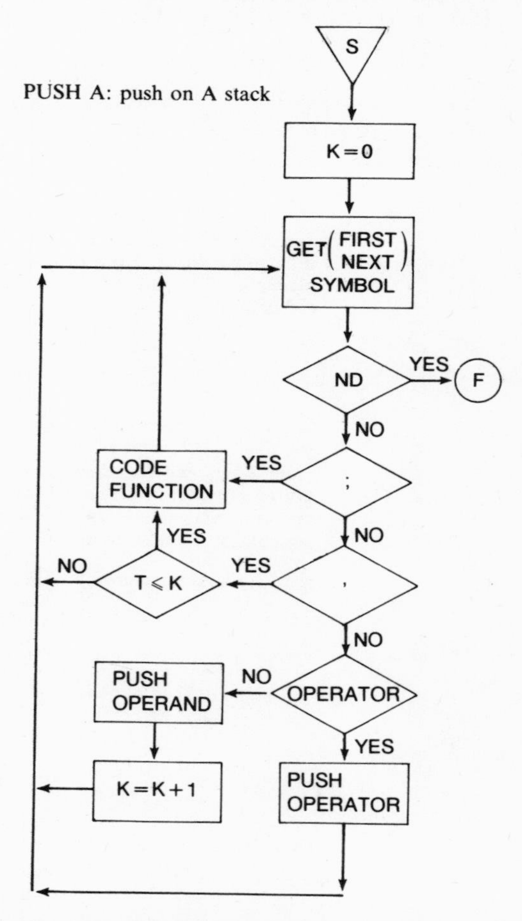

FIGURE 7-6 Revised General Flow Chart for the Stack Function

Table 7-7
Definitions of Compiler Variables

VARIABLE NAME	MEANING	INITIAL VALUE	EXPLANATION
BOC	Beginning of object code	420	See discussion just prior to Figure 8-2.
SBOC	Save beginning of object code	BOC	See discussion just prior to Figure 8-9.
BS	Beginning of symbol string	SEC	SEC was used in Figure 4-1 to save the end of the character string.
K	Counts operands in operand stack	O	This variable will be re-initialized in the code function.
RAN	Top of operand stack	ETL+1	When the algorithm in Figure 5-5 is finished, the value of ETL will be the last location in the operand location table. We will locate the operand push-down stack in the next location.
RAT	Top of operator stack	ETL+11	Here the assumption is that the operand stack will never require more than 10 locations for the operand symbols and the markers between them.
TTL	Transfer table location	ETL+20	See discussion following figure 8-6.
SATL	Statement address table location	0	See discussion just prior to Figure 8-9.

the symbol string which should contain a marker. If it does not, we will transfer to the error function. If it does, we will increment BS by one and check for the character N. If $\overline{BS}=21$, we will increment BS by one and check for the character D. If $\overline{BS}=12$, we are finished with the first pass of the compilation. The initialization described in Table 7-7 and the check for ND are charted in Figure 7-7.

The reader should note carefully the purpose of the ND in the PL/W syntax. Like the END statement in FORTRAN it is used to denote the end of the coding, not the end of the computation. It is, in fact, the symbol that terminates the entire compilation process. Once it is detected we are ready to submit the compiled code to the second pass and then to an optimization routine, or to the operating system where it will be executed.

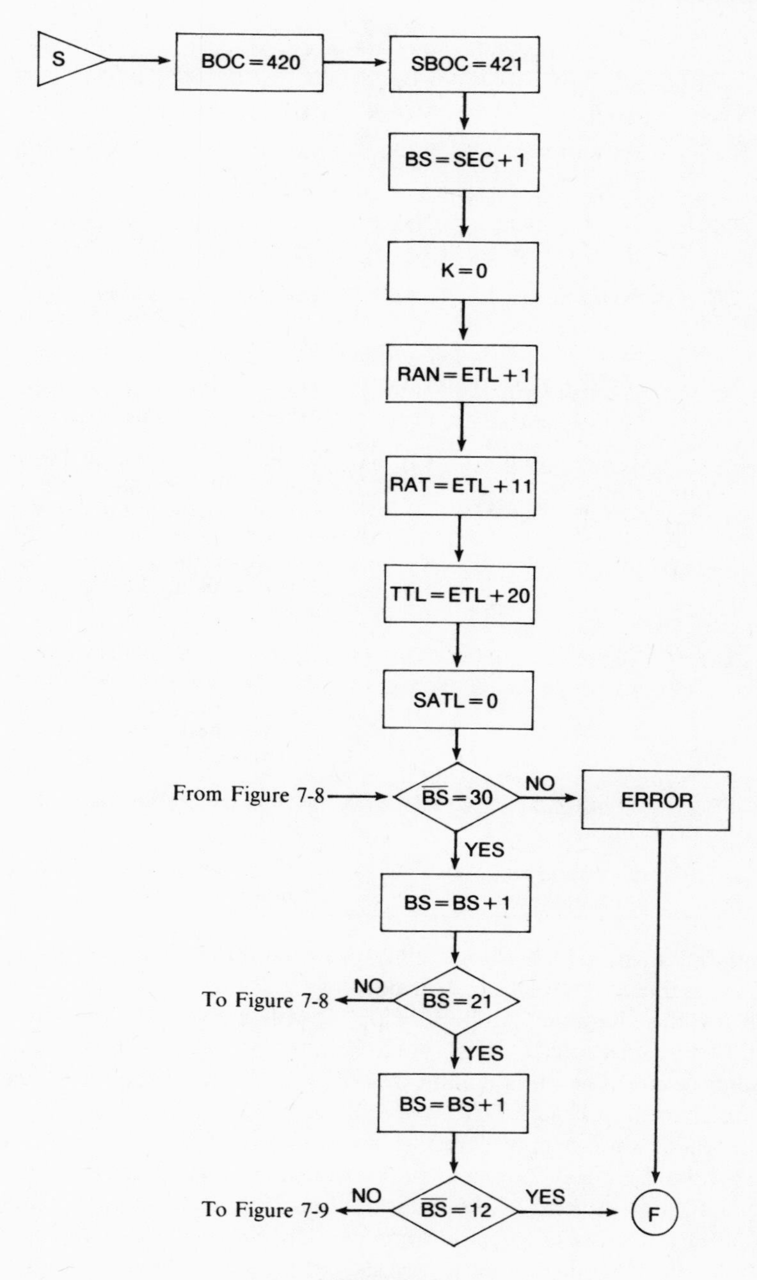
S
BOC = 420
SBOC = 421
BS = SEC + 1
K = 0
RAN = ETL + 1
RAT = ETL + 11
TTL = ETL + 20
SATL = 0
From Figure 7-8
$\overline{BS}$ = 30
NO
ERROR
YES
BS = BS + 1
To Figure 7-8
NO
$\overline{BS}$ = 21
YES
BS = BS + 1
To Figure 7-9
NO
$\overline{BS}$ = 12
YES
F

FIGURE 7-7

If $\overline{BS} \neq 21$, we will check for the character ";." If $\overline{BS} = 23$, control will be transferred to the code function. The details of the code function will be supplied in a later section. When the code function has generated the appropriate code we will increment BS by one and check once again for the marker. If $\overline{BS} \neq 23$, we will check for the character ",." If $\overline{BS} = 22$, we must obtain the type number for the operator on the "top" of the operator stack. This will be done by a table look-up procedure which will obtain the value of T from the operator type table given in Table 5-6. In other words we will implement the type number stack alluded to in Tables 7-2, 7-3, 7-4 and 7-5 by using a table look-up procedure. This procedure will be discussed in section 7-4. Until then, let us continue with our detailed flow chart for Figure 7-6. The treatment of ";" and "," which we have just described is provided in Figure 7-8.

Returning to the check for $\overline{BS} = 12$ in Figure 7-7 we may follow up on what happens when $\overline{BS}$ is a variable. To do this we will have to describe the implementation of a stack.

One of the advantages of a pushdown stack is that the top element is always in the same location. There are several techniques for implementing such a structure in a high-level programming language. Such things as linked lists are very convenient in this regard.

In the WHYCO, however, we will give up the advantage of having the location of the top of a stack remain constant, and instead implement the stack as a list with the "top" of the stack designated as the last element on the list. The variables RAN and RAT can be used to obtain the "top" and should be decremented every time the stack is "popped." Because the POP

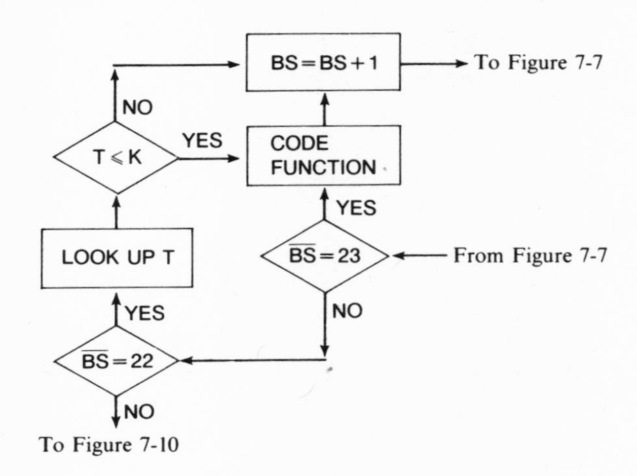

FIGURE 7-8

function will be part of the code function we will, for the moment, only remind the reader that "popping" the stacks will require appropriate manipulation of the marker as well as the operands and operators.

The stack implementation technique that we have just described is reflected in Figure 7-9. In Figure 7-8 we concluded with a test for $\overline{\text{BS}} = 12$. This test will be followed by a test for $\overline{\text{BS}} < 12$ in order to determine if BS is pointing to a number in the symbol string. If it is, control will go to Figure 7-9.

If the test in Figure 7-10 for $\overline{\text{BS}} < 12$ fails, we check for an operator and place it on the operator stack. Figure 7-10 should be examined carefully to insure that the handling of the two-character operator symbols is clearly understood.

We may now conclude our discussion and treatment of the PL/W stack function by combining Figures 7-7, 7-8, 7-9, and 7-10 into a single detailed flow chart. The results of this combination are provided in Figure 7-11.

In section 5-1 we selected a portion of the final flow scheme for the operand location table function and proceeded to generate the appropriate code. The reader should be able to continue that practice in regard to

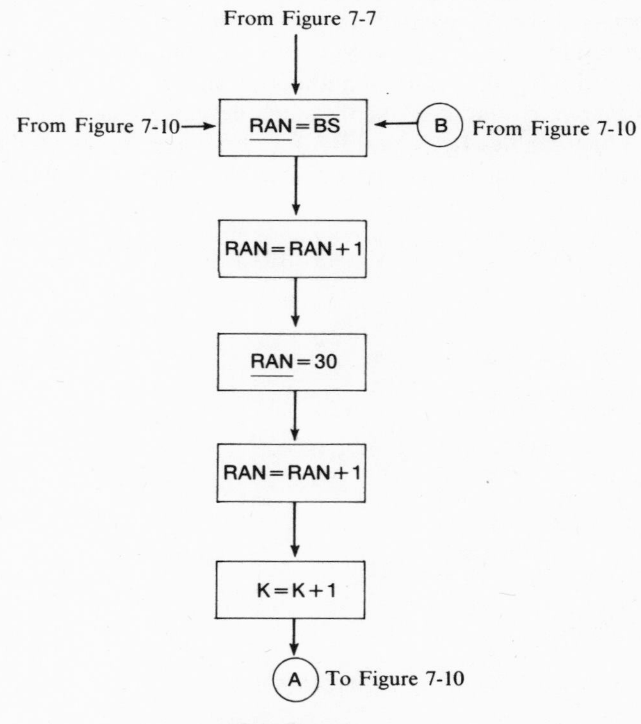

FIGURE 7-9

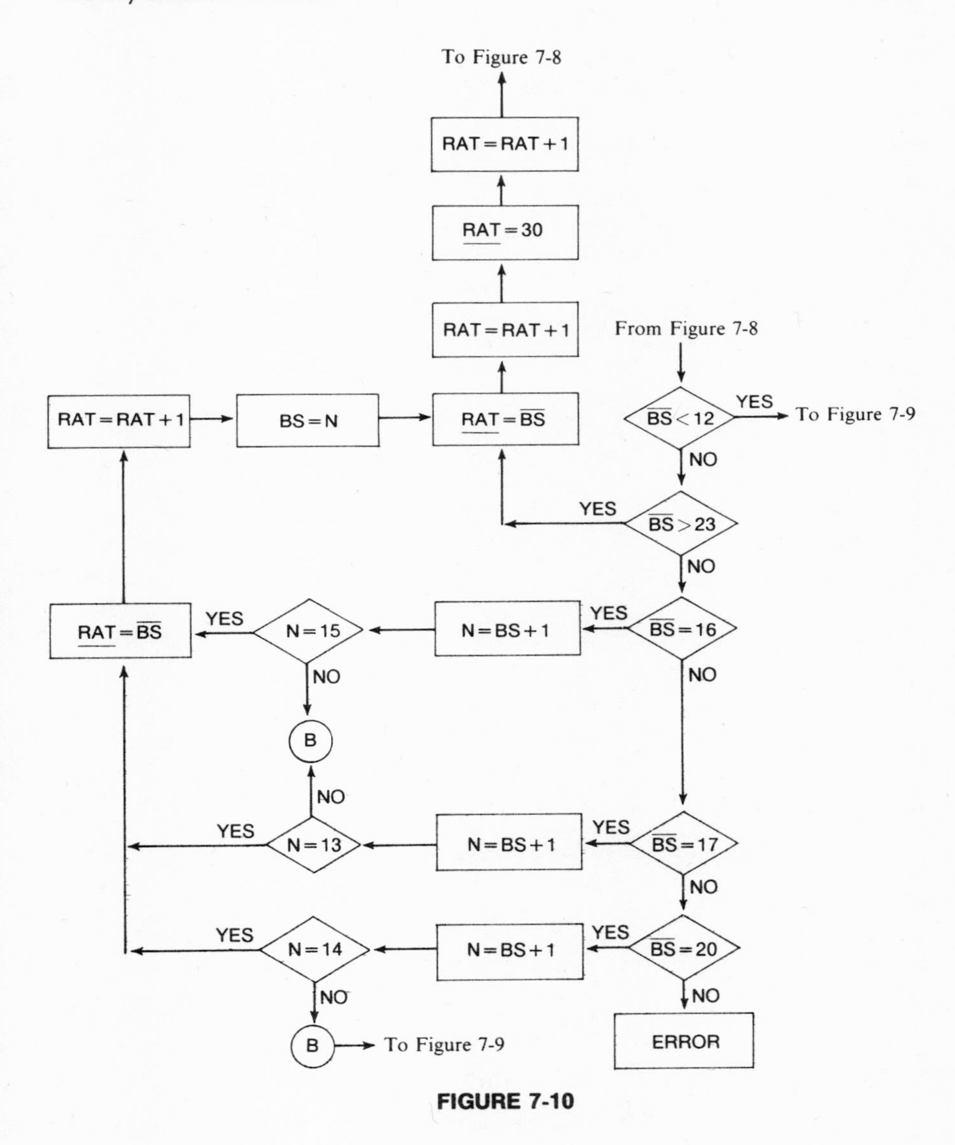

FIGURE 7-10

Figure 7-11. We will not provide it as a formal part of the text because there is nothing significant to be gained by way of conceptual understanding. Individual readers may need more work in this regard and are therefore urged to complete some portion of the code.

We might provide one guideline in regard to the coding. Any statement of the form $\underline{A}=B$ or $\overline{A}<B$ will require one PREP code. Statements of the

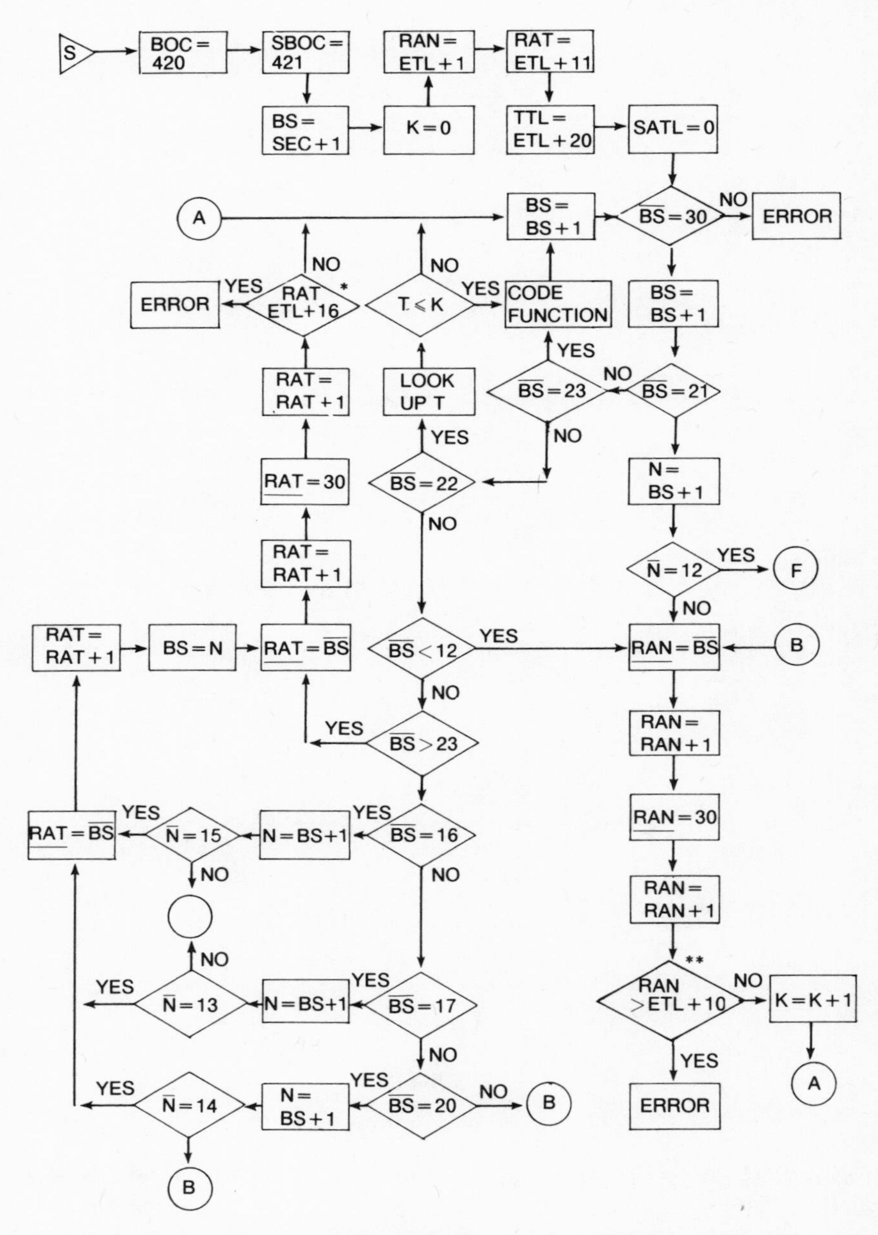

FIGURE 7-11 Algorithm for the Stack Function [*Notes*:*See discussion in Section 8-3 prior to Figure 8-3.** The RAN stack must not be allowed to "erase" the RAT stack.]

form $\underline{A}=\overline{B}$ will require two PREP codes. The best review of these PREP codes can be obtained by examining the coding in Table 4-4.

One final note before considering the implementation of the table look-up. The code function is accessed by the stack function whenever $\overline{BS}=23$ or $\overline{BS}=22$ and $T \leqslant K$. In other words, the code function is part of the stack function and will therefore conclude by returning to the stack function. In Figure 7-11, the reader will also note the addition of two-error conditions which are explained in the notes.

7-4 A Table Look-Up for the PL/W Compiler

The tabling of information is a common feature in all areas of computing and data processing. It is tempting to assume that the procedures for manipulating these tables are simple and straightforward. Actually, just the opposite is the case. Consider, for example, the task of searching for an entry in the table. One could start at the beginning and continue sequentially until the entry is encountered. This is fine for small tables and, in fact, is the procedure we will use in the PL/W compiler. For large tables, however, this would be a very inadequate approach. If the table contained N entries, the sequential search procedure would, on the average, require N/2 comparisons. The logarithmic search consists of entering the table "in the middle" for the first comparison. If the entry is not the one sought, this searching technique goes to the "middle" of the upper half or the "middle" of the lower half for the next comparison. This is continued until the entry is found. Some relatively simple analysis, which the reader can find in numerous sources, shows the maximum number of comparisons for this technique to be $1+\log_2 N$. Graphing $N/2$ and $1+\log_2 N$ will reveal the superiority of the logarithmic search for values of $N>4$.

The reader should be aware that a logarithmic search as well as many other searching techniques requires that the tables have special characteristics. For example, the logarithmic search requires that the entries be indexed and stored in memory in either an increasing or a decreasing order according to the index. Tables are sometimes implemented as linked lists and entries are stored by a technique known as *hashing*. A hashing function converts a key element in the entry to a memory address and thus allows direct access to a table entry. This sounds fine until it is pointed out that the hashing function will not produce unique memory locations for different keys. This is a solvable problem but the programming of the solution is not "simple and straightforward."

Many times the information that is going to be entered in a table requires sorting. Here again we incur an operation, which although relatively easy for a human, is quite complex when implemented in a computer. Sorting techniques have been invented which require special features in the elements being sorted and special data structures to store the results.

Therefore, the reader should not assume that the simplicity of the table look-up procedure for the PL/W compiler is common in other tabling operations. It is, however, very instructive from an introductory point of view.

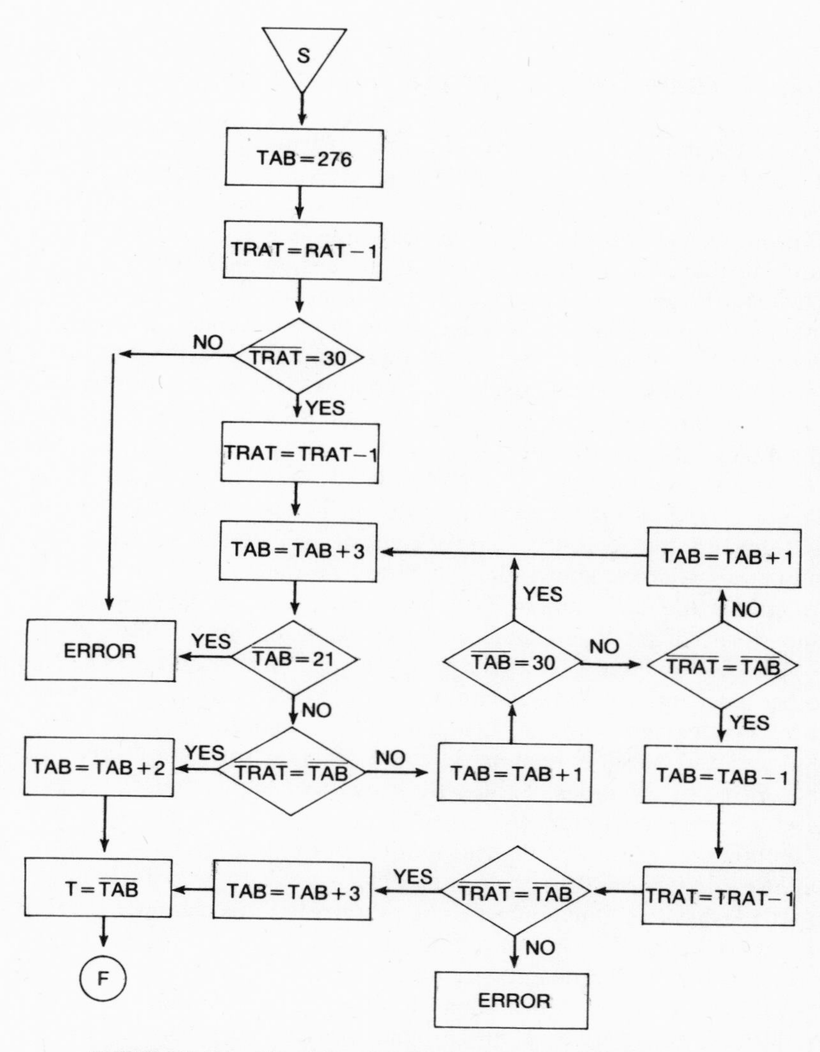

FIGURE 7-12 Algorithm for Table Look-Up Procedure

The parameters in Table 7-8 will be used in implementing the table look-up procedure.

Table 7-8
Definitions of Compiler Variables

PARAMETER	INITIAL VALUE	MEANING
TRAT	RAT	Top operator of operator stack
TAB	276	The incrementation of TAB will require that TAB be initialized to two locations prior to the beginning of the operator-type table
ND	–	Last operator in operator type table

The detailed flow chart for the table look-up procedure is provided in Figure 7-12. Before tracing through this flow chart, the reader should refer back to Table 5-6. The operators in the operator type table are stored with the type number immediately following each operator. All entries in the operator type table, operators as well as type numbers, are separated by the marker. If we search the table and get to the N (21) of ND, we have an error because this would mean that the stack function had incurred an operator that was not on the operator type table. An error of this kind would most likely be due to some malfunction of the hardware. We also incur the ERROR block when agreement on the first character and disagreement on the second character of a two-character operator occurs. The reader will note that even for a sequential search of the operator-type table in the PL/W compiler one could hardly describe the procedure as "simple and straightforward."

EXERCISES

1. Complete Figure 7-2 by drawing the tree structures that were omitted.
2. Complete Table 7-5.
*3. One simplification of the PL/W syntax would be to delete multi-character symbols altogether. This is easily accomplished by using I for IF, G for GT, S for SP, and N for ND. We did not make this simplification because we wanted the reader to have some encounter with multi-character symbols. Revise Figure 7-11 with this simplification in mind.

*This exercise is more on the order of a project and will require a great deal of time and effort. The reader may wish to attempt the solution after the rest of the text has been completed.

Chapter 8

The Code Function

We are now prepared to generate the actual code for a program written in PL/W. In order to motivate this aspect of the PL/W compiler let us rewrite our Fibonacci sequence program in WHYCO code. By so doing we can begin to formulate the code structures which will be necessary for the compiler. We shall designate the memory locations which contain the coded program by a subscripted C. Table 8-1 contains the Fibonacci sequence program in WHYCO code.

There are several important observations which need to be made with respect to Table 8-1.

1. The operations in memory locations C_{23}, C_{24}, and C_{31} contain operand addresses which must be obtained by a special procedure (see section 8-2).
2. The memory locations denoted by "9", "N", "P", "T", "F", and "4" would be obtained from the operand location table.
3. Code sequences such as C_{13}, C_{14}, C_{15} and C_{20}, C_{21}, C_{22} are redundant in the sense that we clear a value from the accumulator and then immediately add it back to the accumulator. The reason for this will become apparent as we proceed with our design of the code structures in (5).
4. "TEMP" is a temporary storage location which must be specified in the PL/W compiler.
5. Each operation is represented by a code pattern which will provide the key to the design of the code structures used in the compiler.

Let us deal with these observations in reverse order.

Table 8-1
WHYCO **Code for Fibonacci Sequence**

MEMORY LOCATION:	CODE	SOURCE CODE
C_1:	CLA	N=9;
C_2:	ADD "9"	
C_3:	STA "N"	
C_4:	CLA	P=1;
C_5:	ADD "1"	
C_6:	STA "P"	
C_7:	CLA	T=1;
C_8:	ADD "1"	
C_9:	STA "T"	
C_{10}:	CLA	FP+T;
C_{11}:	ADD "T"	
C_{12}:	ADD "P"	
C_{13}:	STA "TEMP"	
C_{14}:	CLA	
C_{15}:	ADD "TEMP"	
C_{16}:	STA "F"	
C_{17}:	CLA	IF, F−N, 1, 4;
C_{18}:	SUB "N"	
C_{19}:	ADD "F"	
C_{20}:	STA "TEMP"	
C_{21}:	CLA	
C_{22}:	ADD "TEMP"	
C_{23}:	TRN C_{25}	
C_{24}:	TRA C_{32}	
C_{25}:	CLA	P=T;
C_{26}:	ADD "T"	
C_{27}:	STA "P"	
C_{28}:	CLA	T=F;
C_{29}:	ADD "F"	
C_{30}:	STA "T"	
C_{31}:	TRA C_{10}	GT, $\underline{U}$4;
C_{32}:	STP	SP;

8-1 Code Structures for PL/W

In regard to (5) we can make up a table of code structures for each of the PL/W operators. These structures will be stored in memory with zeros in the operand address portion of the word. When the top of the operator stack is "popped" we will do the following:

1. Obtain the appropriate code structure for the current operator.
2. Obtain the type number (T) from the operator-type table.
3. Pop the operand stack T times.
4. Look up the locations for each operand obtained in (3) in the operand location table.
5. Add the operand locations into the code structure.
6. Store the code structure in memory as object code.

As we proceed we will continually formalize the above procedure until we finally reach a point where a detailed flow chart is possible. In Table 8-2 we have provided a code structure for each operator in its mnemonic and octal form. We remind the reader that we are still using only a nine-bit word in our coding, whereas our word length will have to be considerably longer than that in order to accommodate the increase in memory which will be necessary for storing the compiler. This increase will not affect the first three bits which contain the operation code. Consequently, a word such as 500 should be interpreted as 50...0.

Table 8-2
Code Structures for PL/W Operators

Operator	Mnemonic Code Structure	Octal Code Structure
U̲	CLA 00	600
	SUB 00	100
	STA 00 ("TEMP")	500
−	CLA 00	600
	SUB 00	100
	ADD 00	000
	STA 00 ("TEMP")	500
+	CLA 00	600
	ADD 00	000
	ADD 00	000
	STA 00 ("TEMP")	500
=	CLA 00	600
	ADD 00	000
	STA 00	500
GT	TRA 00	300
IF	CLA 00	600
	ADD 00	000
	TRN 00	400
	TRA 00	300
SP	STP 00	700

In regard to (4) on p. 121, we shall designate TEMP as location 00, and point out that when a value is stored in TEMP it is actually placed back on the "top" of the operand stack (see comments preceding Figure 7-5).

In regard to (3) on p. 121, we will make only a brief comment. Often the code generation will produce code sequences that do such things as storing a value, clearing the accumulator, and then, adding to the accumulator the value that was just cleared from the accumulator. The most vivid example of this in PL/W is the assignment statement. P+T would be coded as:

```
CLA
ADD "P"
ADD "T"
STA "TEMP"
```

The assignment F=P+T would be coded as:

```
CLA
ADD "TEMP"
STA "F"
```

Clearly the code for the assignment statement can be "optimized" by the following code:

```
CLA
ADD "P"
ADD "T"
STA "F"
```

In some computers there is an optimizing function which detects redundant coding and replaces it with optimized coding. Some recent opinion among computer scientists is that optimizing is not worth the trouble. As far as the PL/W compiler is concerned we will not provide the optimizing function, but merely alert the reader to the fact that optimization is an option that may require some consideration in various compilers.

As far as (2) on p. 121 is concerned we will use a table look-up procedure similar to the one provided in Figure 7-12. The one most important difference lies in the fact that the operands are all single-character symbols and as a result they, along with their locations, do not need to be separated by the marker (see Table 5-6). The table look-up procedure for the operand location table will be designed later.

8-2 Two-Pass Compiling

Item (1) on p. 121 represents a very important feature of our particular compiler. The problem concerns the conditional and transfer statements in the PL/W syntax. When a transfer of control is made in the source code it is either forward or backward. Consider the following segments of two different PL/W programs.

```
        .                          .
        .                          .
        .                          .
 (i)    GT,2;             (i)      GT,2;
 (i+1)  F=P;              (i+1)    F=P+G;
 (i+2)  T=N;              (i+2)    T=N;
        .                          .
        .                          .
        .                          .

 PROGRAM SEGMENT 1        PROGRAM SEGMENT 2
```

When line (i) is executed the transfer of control is to line ($i+2$). However, a transfer of two statements forward in the source code is not a transfer of two statements forward in the object code. Consider the object code for the above listings:

```
 .      .      .                  .      .       .
 .      .      .                  .      .       .
 .      .      .                  .      .       .
 (i)    (j)    TRA (J+4)          (i)    (j)     TRA (J+8)
 (i+1)  (j+1)  CLA                (i+1)  (j+1)   CLA
        (j+2)  ADD "P"                   (j+2)   ADD "G"
        (j+3)  STA "F"                   (j+3)   ADD "P"
 (i+2)  (j+4)  CLA                       (j+4)   STA "TEMP"
        (j+5)  ADD "N"                   (j+5)   CLA
        (j+6)  STA "T"                   (j+6)   ADD "TEMP"
 .      .      .                         (j+7)   STA "F"
 .      .      .                  (i+2)  (j+8)   CLA
 .      .      .                         (j+9)   ADD "N"
                                         (j+10)  STA "T"
                                  .      .       .
                                  .      .       .
                                  .      .       .

 PROGRAM SEGMENT 1                PROGRAM SEGMENT 2
```

Note that in both program segments the transfer in the source code was from (i) to ($i+2$), whereas in the object code the transfer in program segment 1 was from (j) to $(j+4)$ and in program segment 2 it was from (j) to $(j+8)$. In other words, there is no way to directly relate the transfer of control in the source code to the corresponding transfer of control in the object code.

In order to solve this problem we shall use a technique known as two-pass compiling. Consider a set of "empty" memory locations which will contain the final version of the object code for some particular PL/W program. During the first "pass" over these locations the code function will fill them with object code which is complete except for the memory address portion of all words whose operator portion is either TRA or TRN. During this pass the code function will create two tables, the **statement address table** and the **transfer table**.

The statement address table will consist of two entries for every source statement. The first entry will be the number of the source statement and the second entry will be the location of the first line of object code for that source statement. The transfer table will also contain two entries. The first entry will be the object code address of any object code containing TRA or TRN. The second entry will be the source statement to which the transfer is being made. Using Table 8-1 as a guide we may illustrate the statement address table and the transfer table after the first pass. This is done in Tables 8-3 and 8-4.

The right-hand entry in the transfer table may require some explanation. The code function will count the source statements as they are processed. This is easily accomplished because of the occurrence of ";" at the end of each statement. For example, when the IF statement in the Fibonacci sequence is processed, the count will be at 5. The operands of the IF can then be added to the 5 to produce the 6 and 9. In the case of the GT statement the count will be 8 and the subtraction of the operand will produce the 4.

The second "pass" consists of looking up each right-hand entry of the transfer table in the left-hand entry of the statement address table. Then,

Table 8-3
Statement Address Table

SOURCE STATEMENT NUMBER	OBJECT CODE LOCATION
1 N=9;	C_1
2 P=1;	C_4
3 T=1;	C_7
4 F=P+T;	C_{10}
5 IF,F−N,1,4;	C_{17}
6 P=T;	C_{25}
7 T=F;	C_{28}
8 GT, $\underline{U}$4;	C_{31}
9 SP;	C_{32}

Table 8-4
Transfer Table

Object Code Address of TRA and TRN	Source Statement Transferred to
C_{23}	6
C_{24}	9
C_{31}	4

the corresponding right-hand entry in the statement address table can be added into the location specified by the corresponding left-hand entry of the transfer table.

It would be possible to handle all "backward" transfers during the first pass because the statement address table would contain the necessary information. This might be implemented as an improvement but transfers "forward" would have to be delayed until the source statement referred to was processed. In the case of the PL/W compiler we will handle all transfer coding after all the statements have been processed.

8-3 The PL/W Code Function

We now turn our attention to the design of the algorithms which will accomplish the six tasks enumerated in the list on p. 121. The data structures we will require are as follows:

1. Operator Type Table
2. Operand Location Table
3. Operator Stack
4. Operand Stack
5. Code Structures Table

We have all of these except the last one. Let us assume that it is loaded into the machine together with the rest of the compiler and is stored in the location that follows the end of the operator type table. We have assumed that the operator type table would begin in location 300. Referring to Table 5-6 reveals that the end of the operator type table occurred in 344. We will, therefore, load the code structures table in the memory locations beginning with 345.

In Table 8-2 we have indicated the contents of the code structures table. However, we must be very specific as to how we are going to provide access to the desired structure. We will accomplish this by placing each operator in the table followed immediately with the corresponding code structure. The marker can be used to separate the operators and code structures. The code structures table is given in Table 8-5. The content of each location is given in its octal form as provided in Table 8-2.

Table 8-5
Code Structure Table

LOCATION:	CONTENT	EXPLANATION
345:	30	marker
346:	24	ICC for U̲
347:	30	marker
350:	600	code structure for U̲
351:	100	
352:	500	
353:	30	marker
354:	25	ICC for −
355:	30	marker
356:	600	code structure for −
357:	100	
360:	000	
361:	500	
362:	30	marker
363:	26	ICC for +
364:	30	marker
365:	600	code structure for +
366:	000	
367:	000	
370:	500	
371:	30	marker
372:	27	ICC for =
373:	30	marker
374:	600	code structure for =
375:	000	
376:	500	
377:	30	marker
400:	16	ICC for G
401:	15	ICC for T
402:	30	marker
403:	300	code structure for GT
404:	30	marker
405:	17	ICC for I
406:	13	ICC for F
407:	30	marker
410:	600	code structure for IF
411:	000	
412:	400	
413:	300	
414:	30	marker
415:	20	ICC for S
416:	14	ICC for P
417:	30	marker
420:	700	code structure for SP

We now have all the data structures required for implementing the PL/W code function.

Continuing in the listing on p. 121, we note that there is no provision in the WHYCO operators that will allow for subroutines. Therefore we cannot take advantage of the table look-up algorithm in Figure 7-12 which has already been coded as part of the stack function. Consequently, we will have to recode it in the code function.

There are other look-up procedures that will also be required in the code function. We will use the operator on the "top" of the operator stack to look up the appropriate code structures. Also, we must look up the locations of the top element(s) of the operand stack in the operand location table.

In addition to the look-up procedures we must PUSH TEMP onto the operand stack whenever the operator is $\underline{U}$, $\overline{-}$, or + and POP the operand stack a total of T times where T is the type number of the operator for which code is being generated. These operands are associated with their corresponding locations in memory and these locations must be combined with the code structure to obtain the object code. This "combining" will be done in a special set of working locations.

And finally, we will have to make entries at the appropriate times in the transfer table and the statement address table. These procedures are summarized in the general flow chart in Figure 8-1. In order to keep the flow chart from becoming cumbersome we provide the following key:

A: Increment source statement counter when $\overline{BS}=23$
B: Increment object code location
C: POP the operator stack
D: Set directional indicators or look up code structure
E: Store code structure in work space
F: Look up T in operator type table
G: POP the operand stack
H: Look up location in operand location table
I: Make appropriate entries in the transfer table
J: Enter location in appropriate section of code structure
K: PUSH TEMP on operand stack
L: Make appropriate entries in statement address table
M: Load object code

The numbers enclosed in parentheses in Figure 8-1 are references to the six steps listed at the beginning of Section 8-1 and are provided so that the reader may compare our initial conceptualization of the code function with our current conceptualization of the code function.

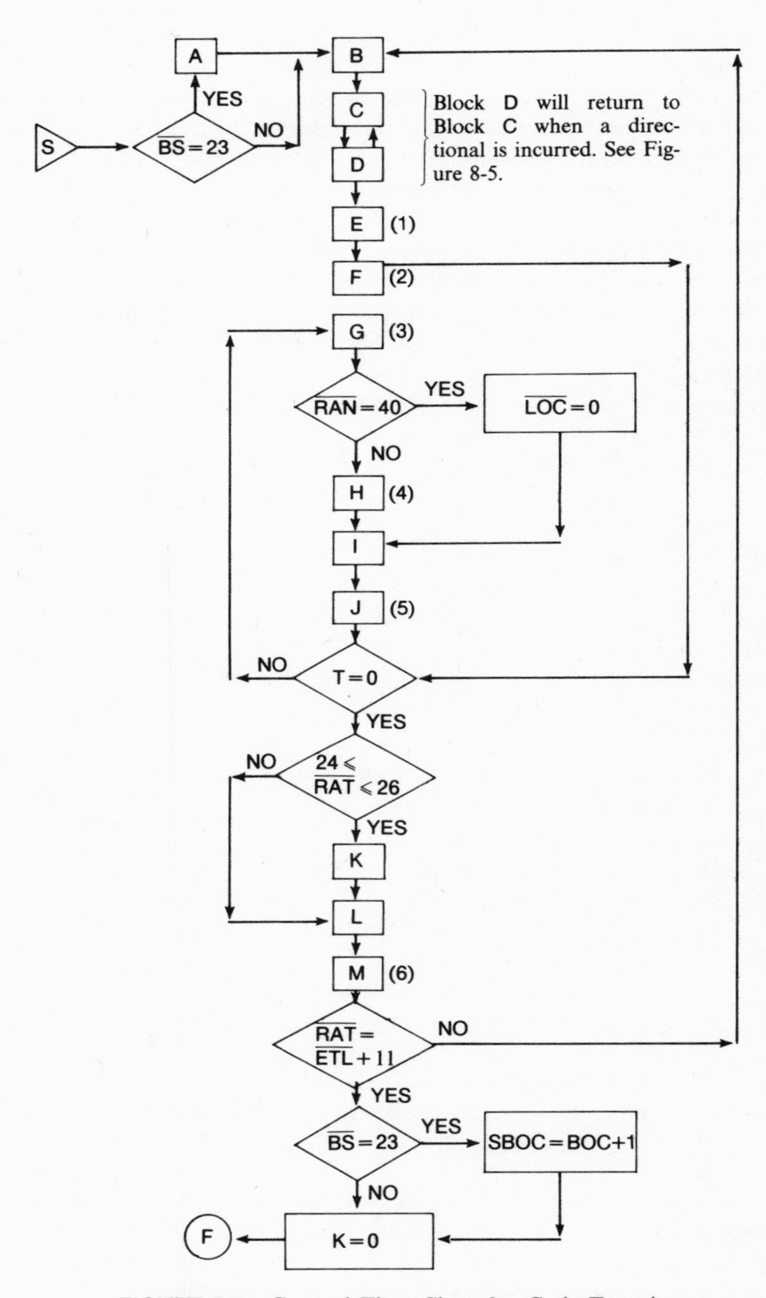

FIGURE 8-1 General Flow Chart for Code Function

As in previous sections, our approach will now be to convert the general flow chart in Figure 8-1 into a detailed flow chart for the PL/W code function. Because of the complexity of this final section of the PL/W compiler, we will develop individual portions of the code function just as we did with the table and stack functions. The reader may then assemble them into a final flow chart. When this is completed we have only the task of completing the object code by filling in the memory address portions of the object code which contain TRA or TRN in the operator portion.

Blocks A and B can be handled in the following manner. Assume a location for the source statement counter is designated by SSC. This location will be loaded with a zero when the compiler is loaded. The incrementation can then be accomplished in the standard way, namely,

SSC = SSC + 1

The object code can be stored anywhere, but a natural location would be immediately following the code structure table. The code structure table ends with location 420. Consequently, we shall use the compiler variable BOC for the beginning of the object code and initialize it to 420 as we did in Figure 7-11. We can then increment it in the following manner:

BOC = BOC + 1

Block C requires that we decrement RAT twice and check the first time for $\overline{\text{RAT}}=30$ and the second time for $\overline{\text{RAT}}>23$. If $\overline{\text{RAT}}>23$, we can proceed to the portion of block D which checks for the directionals. If directionals are incurred, we will set the appropriate indicators in block D and then return to block C of Figure 8-1. If we do not incur a directional, we may proceed with the task of locating the appropriate code structure for $\overline{\text{RAT}}$. If $\overline{\text{RAT}}$ is not greater than 23, we may assume that RAT is pointing to the second character of one of the two-character operators. We will therefore decrement RAT by one and then proceed to transfer to the part of the look-up procedure for the code structures that look for GT, IF, and SP. This can be accomplished by creating a compiler variable for the beginning of the code structure table. We will denote this variable by BCS. When $\overline{\text{RAT}}>23$ we will set BCS=355 and when $\overline{\text{RAT}}$ is not >23 and <30 we will set BCS=377. Figure 8-2 is for blocks A, B, and C of Figure 8-1.

In block D we must deal with the directional operators. When these occur we will set a pointer (D) to locations 276 or 277 and insert a "1" or "0" in those locations in order to indicate "↑" and "↓" respectively. The

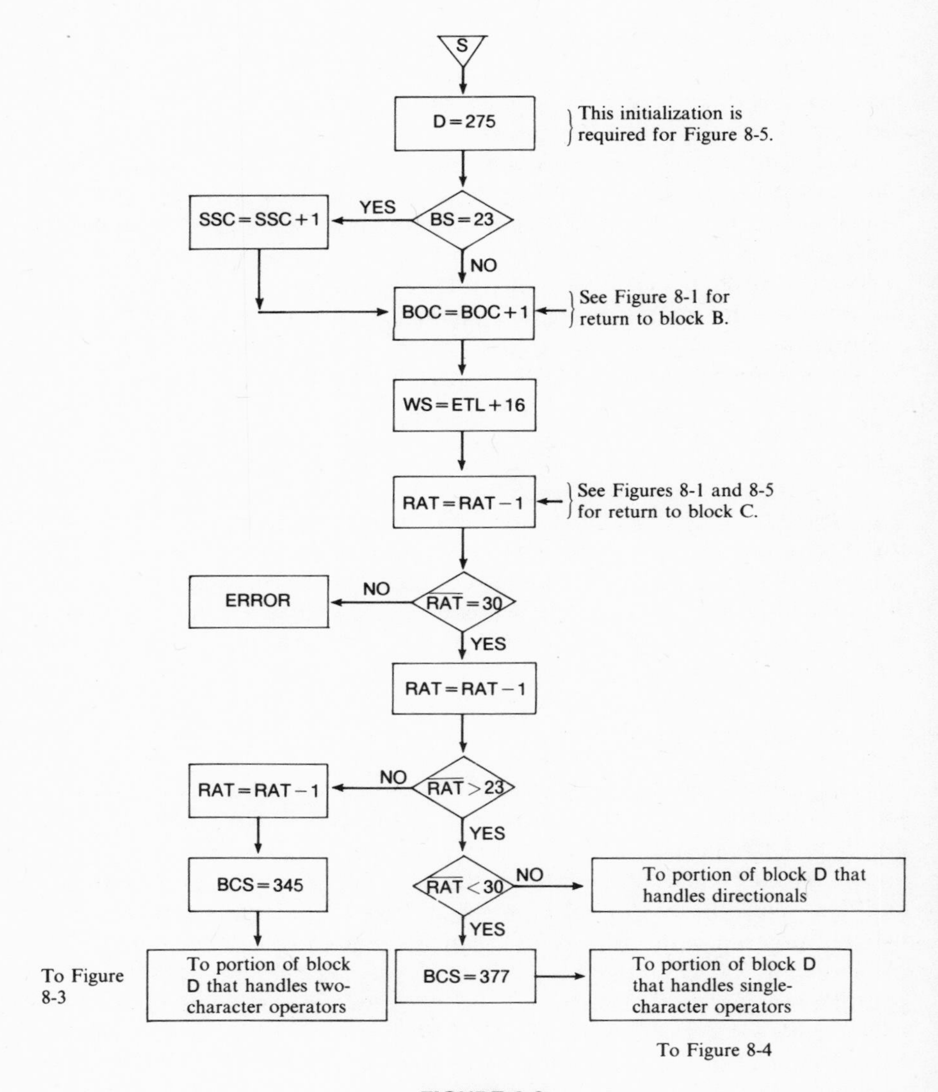

FIGURE 8-2

usefulness of these insertions will become apparent in Figure 8-8. Block D must also search the code structure table for the value of RAT. When this value is incurred the code structure associated with it must be stored in the work space. The reader will recall that the "push-down" stack operators began in location ETL+11, where ETL was the last location in the operand location table. The reader may also examine Table 8-5 to determine the maximum length for any code structure. For the PL/W operators, the maximum length is 4. Therefore, we will designate ETL + 16, ETL + 17, ETL + 18, and ETL + 19 as the work space locations. Here we are assuming that the operator stack will never require more than five locations.

In Figure 8-2 we show three paths to block D. Let us pursue the one associated with BCS = 377. Here we are looking for a double-character operator and should locate it prior to location 417. If we do not locate it prior to 417, we will transfer to the error function. If we do, we will go to block E. This portion of block D is detailed in Figure 8-3.

For the case when BCS = 345 we will follow a similar approach. Here, however, we are looking for a single-character operator and should locate it prior to location 373. Again, if we do not locate it prior to 373, we will

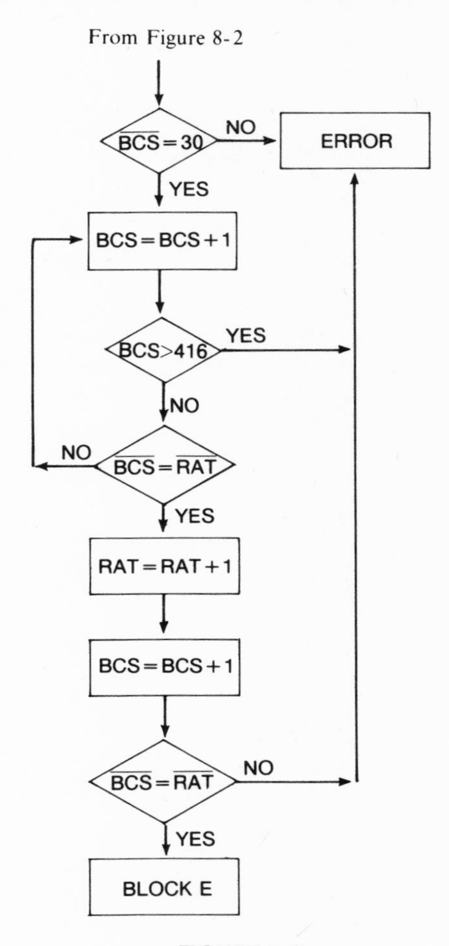

FIGURE 8-3

transfer to the error function. If we do, we will go to block E. This portion of block D is detailed in Figure 8-4.

And finally, we must deal with the occurrence of the directionals on the operator stack. The reader will recall that the directionals are pseudo-operators in the sense that their occurrence does not cause the production of object code. Instead, they set values of "0" or "1" in both of the locations 276 and 277 when we are processing an IF, and only in location 276 when we are processing a GT. Figure 8-5 details the procedure.

For block E we must begin by initializing a counter (COC) which will count the number of object code lines loaded into the work space. We will then load the work space, which was initialized in Figure 8-2 with the appropriate code structure. This process is detailed in Figure 8-6.

When the code structure has been entered into the work space, we will incur the marker. Here we will go to block F. However, if we use Figure 7-12 for the table look-up, we must increment RAT by two upon entering block F and decrement it by two after exiting block F. These incrementations and decrementations are because of the treatment of RAT in Figure 8-2. Their implementation is shown in Figure 8-6.

Also, before entering block G we must check the value of T. Because SP is a null operator, we must not attempt to obtain operands for it. The reader will note the flow from block F to block G in Figure 8-1.

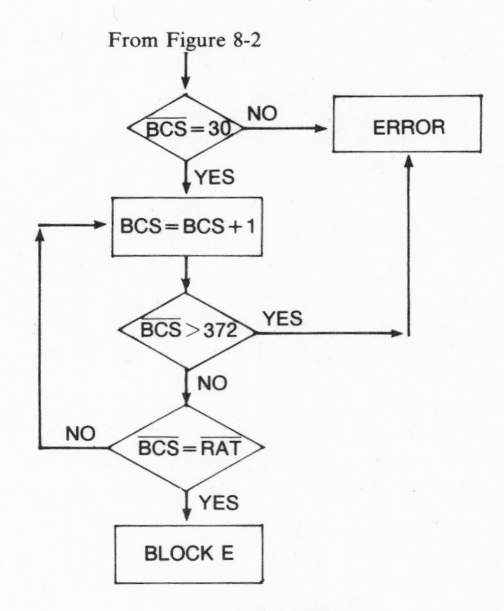

FIGURE 8-4

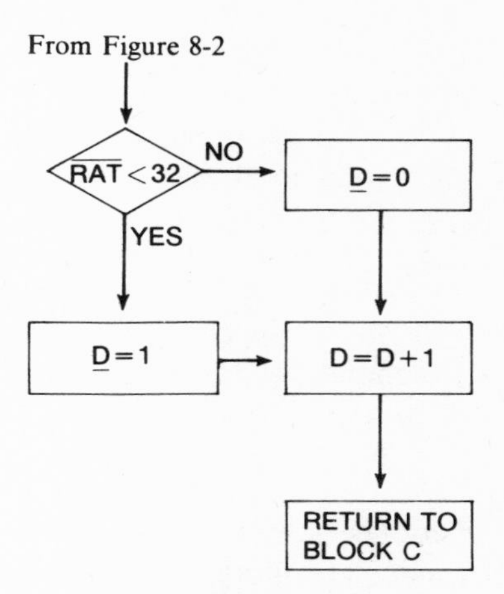

FIGURE 8-5

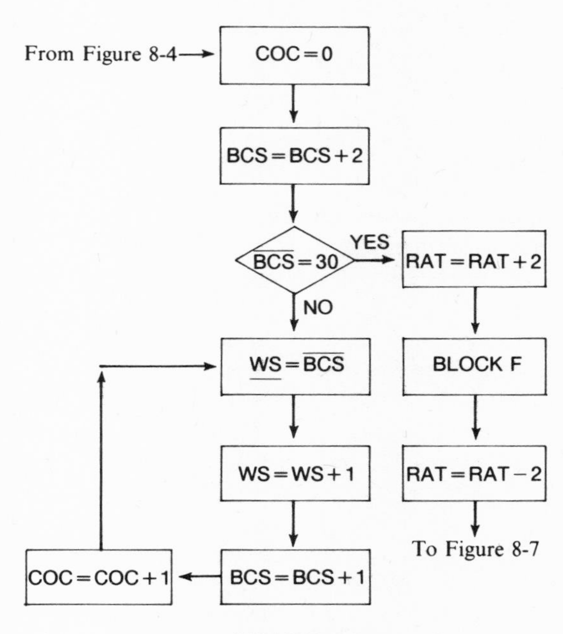

FIGURE 8-6

Block G initializes TAB, decreases RAN by one, and then checks for the marker. If the marker is incurred, we decrement RAN and continue. In the check for $\overline{\text{RAN}}=\overline{\text{TAB}}$ we could incur the variable TEMP on the "top" of the operand stack. We will designate an ICC of 40 for TEMP. Let us deal now with the operands other than TEMP.

In block H we must look up the operand's location in the operand location table. The reader may refer to Table 5-2 for a sample operand table and to Figure 5-5 for the table construction algorithm. There are three items of particular significance for block H. First, recall that the operand's location is the next location in the operand location table following the ICC entry. Second, the variable STL contains the beginning location for the operand location table. And third, the operand location table ends with the marker. The table look-up is quite simple because of the single-character operands. Therefore, we present blocks G and H in Figure 8-7 without further discussion.

Once we have the location we must check to see if the current operator is GT or IF. If it is, we must make an appropriate entry in the transfer table. Here we will take advantage of the simplicity of the PL/W operators and assume that $\overline{\text{RAT}}$=T(i.e. ICC=15) means we are processing a GT and $\overline{\text{RAT}}$=F(i.e. ICC=13) means we are processing an IF. We will not back up RAT by one and check for the G of GT or the I of IF. The reader should note the way the two-character operators are processed in Figures 8-2 and 8-3. In Figure 8-2 we did back up RAT to check for the G of GT, the I of IF, and the S of SP. In the event that both characters of these two-character operators were not incurred in Figure 8-3, we went to the error function. Consequently, it would be redundant to back up RAT in the implementation of block I.

Before we examine the process by which an entry is made in the transfer table we must define the transfer table itself. In Table 8-4 we provided an example transfer table. It consists of entries which are ordered pairs. The first element of each pair is the object code address of TRA or TRN and the second element of each pair is the source statement to which the transfer is to be made. The work space was located in ETL+16, ETL+17, ETL+18, and ETL+19. Therefore, we can begin the transfer table in ETL+20. The first ordered pair in the table will be located in ETL+20 and ETL+21, the next ordered pair will be in ETL+22 and ETL+23, and so on. The initialization of the transfer table location (TTL) is given in Figure 7-11.

For the GT operator we have the following code structure:

```
403:   500
```

Consequently, we will require only the first element of the work space, the

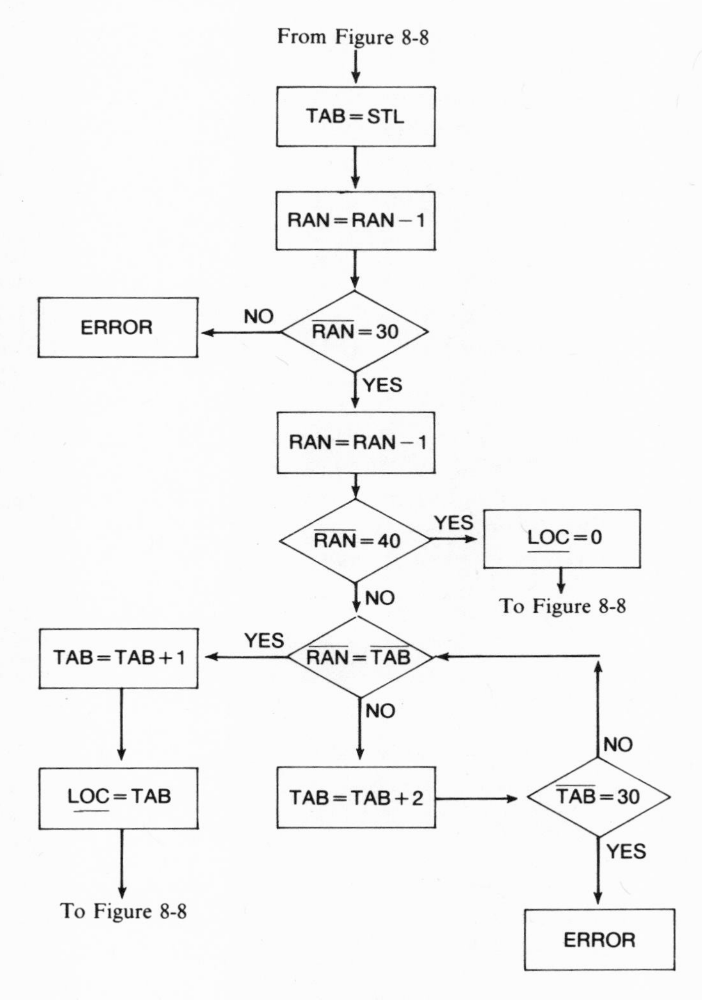

FIGURE 8-7

content of which will be transferred in block J to the location specified by the current value of BOC. Therefore, $\overline{BOC}$ is the first entry of the ordered pair that will go into the transfer table. The second entry will be obtained by combining the value of LOC with the current value of SSC, the source statement counter.

For the IF operator we have the following code structure:

410: 600
411: 000
412: 400
413: 300

Here we require all of the available work space. The transfer commands are in 412 and 413 of the code structure and will therefore be transferred in block J to the locations specified by BOC+2 and BOC+3. Therefore, BOC+2 and BOC+3 are the first entries of two ordered pairs that will go into the transfer table. The corresponding second entries of each pair will be obtained by combining the current value of LOC and SSC. In the event that $\overline{\text{RAT}}$ is neither GT or IF, we proceed directly to block J.

Block J must place the value of LOC into the appropriate part of the code structure. When the current operator is IF or $\underline{\text{u}}$ we must add $\overline{\text{LOC}}$ to the contents of the second location in the work space. When the current operator is −, +, or = we will add $\overline{\text{LOC}}$ to the second location in the work space, pop the operand stack to obtain the second operand, look up the location of the operand ($\overline{\text{LOC}}$), and add $\overline{\text{LOC}}$ to the third location in the work space. Figure 8-8 shows the detailed flow chart for blocks I and J.

Block K returns TEMP(40) to the operand stack whenever the current operator is $\underline{\text{u}}$, −, or +. The reader should recall that the storage of the actual result of the operations $\underline{\text{u}}$, −, and + is done in the code structure.

The implementation of block L requires a definition of the statement address table. Since this table requires an entry for every source statement,

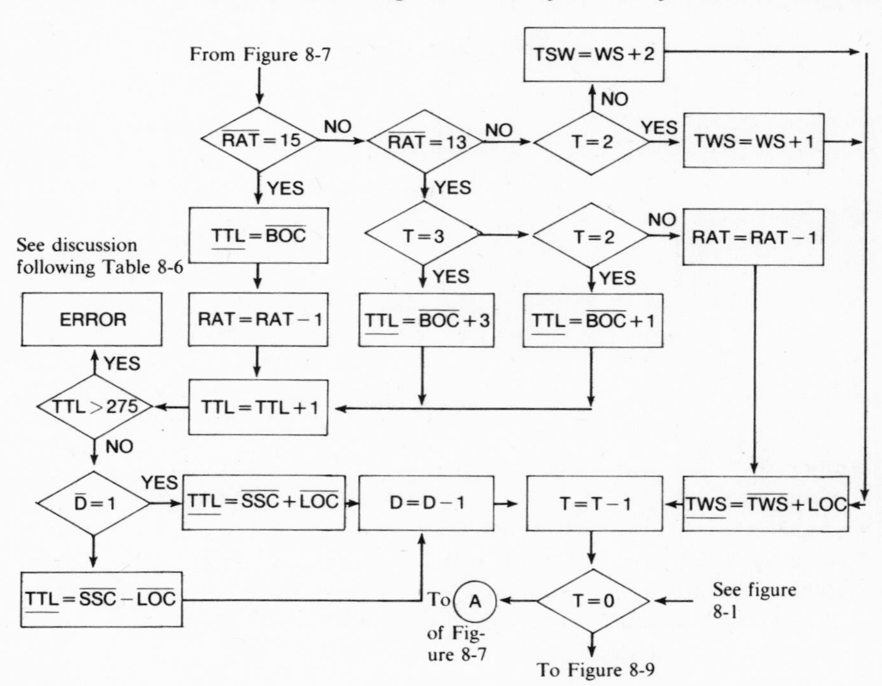

FIGURE 8-8

it will be considerably longer than the transfer table. Also we want to make an entry in the statement address table only when $\overline{BS}=23$, that is, when we have incurred the ";."

The potential length of the statement address table will now provide an excellent opportunity for us to discuss some memory management problems we have previously virtually ignored.

Let us begin this brief digression from the implementation of the code function by summarizing the uses of memory which have been made thus far. As usual, we will use the Fibonacci sequence program to provide specific illustrations. Table 8-6 provides a "memory map" of the situation as it currently stands.

There are several important observations that can be made regarding Table 8-6. First, the operator type table and the code structure table are part of the compiler, whereas all of the other data structures are produced by the compiler. Therefore, if the end of the transfer table goes beyond 275, we have a memory allocation error. Normally, the operating system would take over at this point and perform some kind of reallocation. Because we are not writing an operating system for the WHYCO we will simply transfer control to our error function and stop. In effect, we are telling the user that his or her program is too big.

A similar problem will arise with respect to the object code. Suppose we load the rest of the compiler (i.e., the compiler minus the operator type table and the code structure table) in memory beginning at location 500. Doing so would require that the object code fit in 421 through 477. If it did not, we would incur another memory allocation error. Since 421 through 477 represents 64 memory locations and since the Fibonacci sequence program requires only 35 locations, no such error would occur. However, a program larger than the Fibonacci program could easily cause such an error.

Table 8-6
Memory Map for PL/W Compiler

MEMORY LOCATION (S)	CONTENT
0	Location of TEMP
1 through 60	Character String
61 through 214	Symbol String
234(215 through 233 (ETL)	Operand Location Table
234(ETL+1) through 245(ETL+10)	Operand Stack
246(ETL+11) through 252(ETL+15)	Operator Stack
253(ETL+16) through 256(ETL+19)	Work Space
257(ETL+20) through 275	Transfer Table
276 and 277	Directional Indicators
300 through 344	Operator Type Table
345 through 420	Code Structure Table
421 through ?	Object Code

A second observation involves the length of the stacks. We have already made note of the memory management problem that can arise here (see notes in Figure 7-11).

And finally, a third observation can be made with respect to the character string. For some time now, the character string has not been referenced. In fact, once the symbol string is produced, there is no need to retain the character string. We will take advantage of this fact and reuse locations 1 through 60 for our statement address table. Determining which memory locations are reusable during any computational process is a difficult and sometimes tedious task. Compiling—the computational process with which we are concerned—is one that requires a great deal of memory, especially for high-level programming languages such as FORTRAN, BASIC, and PL/1. Therefore, the skillful management of memory resources is a necessary part of any operating system.

Considering our more immediate task, let us be content with only the above introduction to memory management problems and proceed now to the implementation of block L, the statement address table.

The statement address table can be implemented in the same way as the transfer table, namely, it will consist of ordered pairs. The first element of each pair is the source statement number and the second element of each pair is the object code location where the coding for that particular source statement begins.

Making these entries will require an initialization of a variable that points to the various locations in the statement address table. We will designate this variable as SATL, statement address table location, and will initialize it to zero. This initialization appears in Figure 7-11. The actual entries will be the values of the variables SSC, source statement counter, and BOC, beginning of object code. Consider the case where the code function is entered due to the occurrence of the comma when $T \leq K$. Under these circumstances we will need to save the value of BOC before we load the object code. The reader will recall the IF statement from the Fibonacci sequence program. Here, when the coding for F-N is completed, we will load the code in locations beginning with BOC. The incrementation of BOC during this loading procedure will destroy the value of BOC which contains the beginning location of the object code for the IF statement. We have initialized this value in Figure 7-11 under the variable name SBOC. It is initialized near the end of the algorithm in Figure 8-1.

Block M, which loads the object code, will depend on two variables—BOC and COC. All we need to do is load the work space into the memory beginning at location BOC. The number of incrementations of BOC depends on COC, the variable that counts the number of object code lines loaded into the work space. Figure 8-9 shows the implementation of blocks K, L, and M.

From Figure 8-8

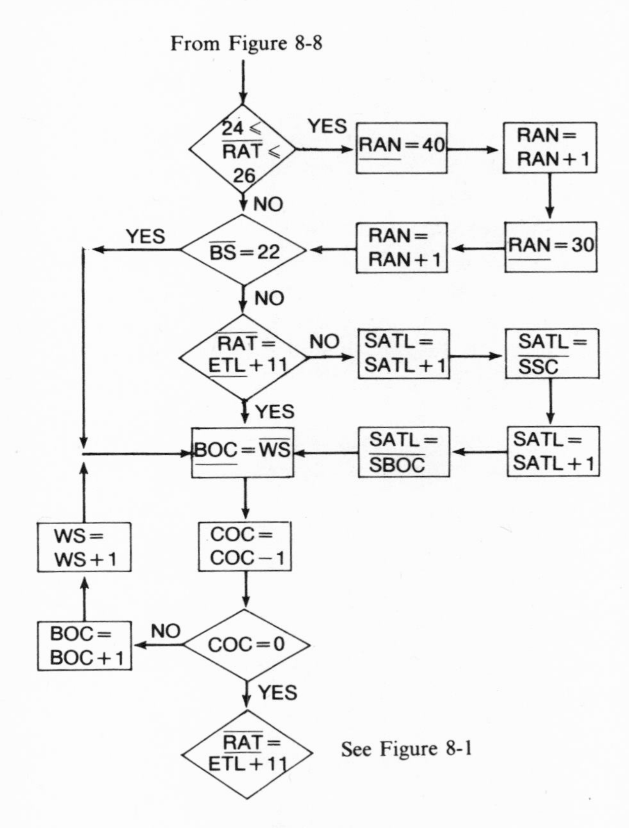

See Figure 8-1

FIGURE 8-9

The figures listed in Table 8-7 represent the complete implementation of the PL/W code function. They are given together with a reference to Figure 8-1.

The only task remaining is to make a second pass over our object code and fill in the memory address locations for the TRA and TRN commands. The algorithm for doing this would be "attached" to Figure 7-11 at F.

Table 8-7

FIGURE	FIGURE 8-1 REFERENCE
8-2	Blocks A, B, C
8-3, 8-4, 8-5	Block D
8-6	Block E
8-7	Blocks G, H
8-8	Blocks I, J
8-9	Blocks K, L, M

Because we know where the transfer table begins, namely, in $\overline{\text{ETL}}+20$, and because the current value of TTL points to the end of the table, let us read the table backwards, decrementing TTL as we go. Doing so will allow us to test TTL against $\overline{\text{ETL}}+20$. When $\overline{\text{TTL}}<\overline{\text{ETL}}+20$, we are finished.

Before we begin the implementation of the second pass algorithm, let us examine the statement address table and the transfer table that appear in Tables 8-3 and 8-4, respectively. In the WHYCO memory they will appear as shown in Tables 8-8 and 8-9.

In these two figures we can observe the possibility of relating the content portion of the transfer table to the memory location portion of the statement address table. In fact, because of the way we have "managed" the memory we could have stored the object code location for each source

Table 8-8
Statement Address Table

MEMORY LOCATION:	CONTENT
1:	1
2:	C_1
3:	2
4:	C_4
5:	3
6:	C_7
7:	4
10:	C_{10}
11:	5
12:	C_{17}
13:	6
14:	C_{25}
15:	7
16:	C_{28}
17:	8
20:	C_{31}
21:	9
22:	C_{35}

Table 8-9
Transfer Table

MEMORY LOCATION:	CONTENT
$\overline{\text{ETL}}+20$:	C_{23}
$\overline{\text{ETL}}+21$:	6
$\overline{\text{ETL}}+22$:	C_{24}
$\overline{\text{ETL}}+23$:	9
$\overline{\text{ETL}}+24$:	C_{34}
(TTL) $\overline{\text{ETL}}+25$:	4

statement in a memory location whose value is the same as the number of the source statement. Other memory management techniques would not have made this possible, however. In our case, the number of the "source statement transferred to," which is the content portion of the transfer table, is found in a memory location that is equal to twice its value less one. Consequently, the "object code location" for the "source statement transferred to" is stored in a memory location whose value is exactly twice the value of the "source statement transferred to." We will take advantage of this in the implementation of the second pass algorithm.

The implementation will actually be quite simple. We will double the current value of TTL and save the result. Because this result is, in effect, a pointer to the statement address table, we will use the variable name PSAT. Next we will decrement TTL by one in order to obtain the object code address of the transfer. Adding $\overline{\text{PSAT}}$ and $\overline{\text{TTL}}$ will produce the desired result. This procedure, together with a test on TTL for termination, is provided in Figure 8-10.

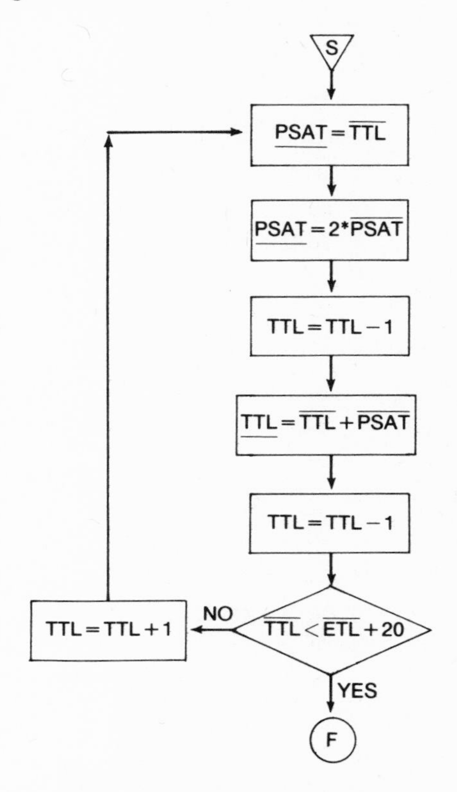

FIGURE 8-10 Second Pass Algorithm

Table 8-10
Coding for the Second Pass Algorithm

MEMORY LOCATION:	CODE	FIGURE 8-10 REFERENCE
F_1:	CLA	PREP CODE FOR F_{10}
F_2:	ADD "000"	
F_3:	ADD "TTL"	
F_4:	STA F_{10}	
F_5:	CLA	PREP CODE FOR F_{11}
F_6:	ADD "500"	
F_7:	ADD "PSAT"	
F_8:	STA F_{11}	
F_9:	CLA	$\underline{\text{PSAT}}\ \overline{\text{TTL}}$
F_{10}:	ADD 00	
F_{11}:	STA 00	
F_{12}:	CLA	PREP CODE FOR F_{21}
F_{13}:	ADD "000"	
F_{14}:	ADD "PSAT"	
F_{15}:	STA F_{21}	
F_{16}:	CLA	PREP CODE FOR F_{23}
F_{17}:	ADD "500"	
F_{18}:	ADD "PSAT"	
F_{19}:	STA F_{23}	
F_{20}:	CLA	$\underline{\text{PSAT}}\ 2^{*}\ \overline{\text{PSAT}}$
F_{21}:	ADD 00	
F_{22}:	SLO	
F_{23}:	STA 00	
F_{24}:	CLA	TTL = TTL − 1
F_{25}:	ADD "TTL"	
F_{26}:	SUB "1"	
F_{27}:	STA "TTL"	
F_{28}:	CLA	PREP CODE FOR F_{41}
F_{29}:	ADD "000"	
F_{30}:	ADD "TTL"	
F_{31}:	STA F_{41}	
F_{30}:	CLA	PREP CODE FOR F_{42}
F_{33}:	ADD "000"	
F_{34}:	ADD "PSAT"	
F_{35}:	STA F_{42}	

Table 8-10 (Continued)

MEMORY LOCATION:	CODE	FIGURE 8-10 REFERENCE
F_{36}:	CLA	PREP CODE FOR F_{43}
F_{37}:	ADD "500"	
F_{38}:	ADD "TTL"	
F_{39}:	STA F_{10}	
F_{40}:	CLA	$\underline{\text{TTL}} = \overline{\text{TTL}} + \overline{\text{PSAT}}$
F_{41}:	ADD 00	
F_{42}:	ADD 00	
F_{43}:	STA 00	
F_{44}:	CLA	TTL = TTL − 1
F_{45}:	ADD "TTL"	
F_{46}:	SUB "1"	
F_{47}:	STA "TTL"	
F_{48}:	CLA	PREP CODE FOR F_{57}
F_{49}:	ADD "100"	
F_{50}:	ADD "ETL"	
F_{51}:	STA F_{57}	
F_{52}:	CLA	PREP CODE FOR F_{59}
F_{53}:	ADD "100"	
F_{54}:	ADD "TTL"	
F_{55}:	STA F_{59}	
F_{56}:	CLA	$\overline{\text{TTL}} < \overline{\text{ETL}} + 20$
F_{57}:	SUB 00	
F_{58}:	SUB "20"	
F_{59}:	ADD 00	
F_{60}:	TRN F_{66}	
F_{61}:	CLA	TTL = TTL + 1
F_{62}:	ADD "TRL"	
F_{63}:	ADD "1"	
F_{64}:	STA "TTL"	
F_{65}:	TRA F_1	
F_{66}:	STP	

This completes the PL/W compiler. However, because Figure 8-10 can be implemented with the use of the SLO operator (recall that shifting one to the left is the same as multiplying by two), an operator we have not used before, and since it has been some time since we generated the actual code for some portion of the compiler, we have completed this section with the WHYCO code for Figure 8-10. The memory locations are designated by a subscript F.

We have completed the hardware and software design for a simple digital computer and an elementary compiler. In this effort we have attempted to reduce all of the conceptual dimensions of these designs to actual implementations. In Chapter 2 we did the following:

1. Related logic design to actual circuitry
2. Developed the concept of subcommand generation
3. Developed the concept of the fetch-execute cycle
4. Designed counters and decoders
5. Designed registers and memory

In Chapters 3 through 8 we did the following:

1. Developed the syntactic approach to language definition
2. Distinguished characters from symbols using internal compiler code and markers
3. Developed tables and table look-up procedures
4. Developed the parsing concept using regular grammar and directed graphs
5. Developed the stack concept and postfix notation using tree structures
6. Studied operator precedence
7. Produced object code using code structures
8. Studied two-pass compiling
9. Developed code for the compiler using the PREP code concept

The reader should now be able to move on to some of the more sophisticated texts on compiling and find that even though the details and concepts are more complicated than those used in this text, they are, for the most part, reducible to the basic concepts set forth above.

One of the most important conclusions which the author has reached while writing such a text concerns the fact that software items such as compilers are best understood by beginning at the "bit" level. Everything we have done is easily reduced to the actual bits in the actual registers of an actual computer. In most texts on compiling, the compiler is written in an implementation language such as PL/1 which, of course, requires its own compiler. This phenomenon was referred to in section 7-2 following Figure 7-3. In these settings, the "implementation language" converts the "source language" into the "target language." There is nothing technically wrong with such an approach. From a pedagogical point of view, however, it disguises a great deal. For example, the simple implementation of a table is almost trivial in most high-level languages and yet, as we have seen in the WHYCO, it can be an arduous task.

Even though there is a great deal left to study in the area of compiler design and implementation, it is the author's experience with students and industrialists alike that during such further study, readers of a text such as this will find themselves constantly referring back to an elementary computer and compiler as a means of conceptualizing what is "going on" in more complex environments.

EXERCISES

1. The variable BOC is used to point to object code locations. It is initialized in Figure 7-11 and incremented in Figure 8-9. Assuming the coding for the compiler starts somewhere beyond the initial value of BOC, insert an appropriate error check in Figure 8-9 to insure that object code does not erase the compiler.
2. The implementation of the "second pass" can actually be simplified by recognizing that there is no need to record the source statement number in the statement address table. Instead, we code and record only the object code locations and the location of the first entry in the table. Let us suppose that the memory location of the first entry is denoted by K. Then the beginning object code location for source statement number L is given by $K+L-1$. Rewrite the second pass algorithm, both flow chart and compiler code, using this method.
3. Consider the following suggestion: Abandon the use of directionals in the GT and IF statements and use the following idea. The $\underline{u}$ will cause a transfer back in the program and the absence of a sign will cause a transfer forward. Also, additional commas will be used to separate these destination numbers. In other words, the IF and GT statements in the Fibonacci sequence program will be altered as follows:

   IF,F-N,↓1↓4; becomes IF,F-N,1,4;
   GT↑4; becomes GT,$\underline{u}$4;

   Assume the appropriate changes have been made to the syntax. Why would this technique fail? For example, what will the code function do with $\underline{u}$4? What happens when an expression occurs in the IF statement and is followed by a negative destination (e.g., IF,T-P,$\underline{u}$2,3;)?
4. As a project, obtain a large sheet of paper, 3 ft by 5 ft for example, and "link up" all the flow charts for the PL/W compiler. This will prove to be more

instructive than it appears. The main flow charts are as follows:

Figure 4-1	Figure 7-12
Figure 5-5	Figures 8-2 through 8-8
Figure 6-5	Figure 8-9
Figure 7-11	Figure 8-10

5. As a project, write a compiler trace for the Fibonacci sequence program. Worksheets such as the one below will assist in organizing the trace. Also, since a trace involves keeping track of every change in the WHYCO memory, consider organizing your worksheets into the following categories:
   a. Character String and Statement Address Table
   b. Symbol String
   c. Operand Location Table
   d. Operand Stack
   e. Operator Stack
   f. Work Space
   g. Transfer Table
   h. Operator Type Table
   i. Code Structures Table
   j. Object Code
   k. Compiler Variables and Constants (these can begin in location 500)

Page______ of______

PL/W Compiler Worksheet
Sample Program:

Notes: ____________________________

____

____

Location	Explanation	Contents

Bibliography

Cleaveland, J. Craig and Robert C. Uzgalis. *Grammars for Programming Languages*. New York: Elsevier North-Holland, 1977.

Flores, Ivan. *The Logic of Computer Arithmetic*. New Jersey: Prentice-Hall, 1963.

Foster, Caxton C. *Computer Architecture*, 2nd ed. New York: Van Nostrand Reinhold, 1976.

Goos, G. and J. Hartmanis, eds. *Compiler Construction: An Advanced Course*. New York: Springer-Verlag, 1976.

Halstead, Maurice H. *A Laboratory Manual for Compiler and Operating System Implementation*. New York: Elsevier North-Holland, 1974.

Maley, Gerald A. and John Earle. *The Logic Design of Transistor Digital Computers*. New Jersey: Prentice-Hall, 1963.

Rosen, Saul. *Programming Systems and Languages*. New York: McGraw-Hill, 1967.

Weingarten, Frederick W. *Translation of Computer Languages*. San Francisco: Holden-Day, 1973.

Williams, Gerald E. *Digital Technology*. Chicago: Science Research Associates, 1977.

Wulf, William, Richard K. Johnsson, Charles B. Weinstock, Steven O. Hobbs, and Charles M. Geschke. *The Design of an Optimizing Compiler*. New York: Elsevier North-Holland, 1975.

Zissos, D. *Problems and Solutions in Logic Design*. London: Oxford, 1976.

Index